Different Differential Equations

Nate Ramer

October 16, 2025

Published by Amazon KDP

First edition published 2025.

WHY? How do I write? Am I asking you this question, or am I asking myself this question? Why am I asking so many questions? Looking introspectively, is it money? Not really. Is it fame? Not particularly. However, one thing struck me: making a difference through passion.

I am passionate about writing, learning, and expressing that in the books I write. I love conveying knowledge in a form that is easily (hopefully) understood; if not by everyone, then by some.

I want people to pick up my book and say that they learned something or that I influenced them in some way. I want to make a difference in people's lives. That may not be achieved by proving the Riemann hypothesis or unifying quantum and relativistic physics. But I will be damned if I don't inspire one person to pursue more information about a topic that they find interesting. I want to be the stepping stone that the people who make these grand proofs and theories use and reference as they begin their mathematical journey, much like I did at one point.

Previously, I wrote Integral Integrals and Some Sums"and"On Rates of Change, a personal favorite that does not get as much attention as it deserves compared to the others. Despite this, I originally wrote it because I was passionate about the topic I was researching. Integrals were something I loved to sit down and try to evaluate, even though I knew that I lacked the fundamental skills to do so. Problem-solving in these instances is highly valuable for the future.

My story started when I was 14 years old; at the time, I was in 8th grade and taking Algebra 2. All I remember from that class was learning about matrices, factoring, and exponential functions. Up until this point, I have excelled in school, but I only did what was expected of me (minus one exception when I had my teacher explain square roots to me when I was 10).I also had an obsession with watching YouTube videos for hours, rotting away my brain, click by click. I loved playing my Xbox for hours at a time while enjoying different videos in the background; Minecraft, Fortnite, you name it; I watched it. As a kid, I never thought that I would have a passion for math, science, and even reading. In 5th grade, as part of my accelerated program, I was placed into a pre-algebra class. When I say this, I say it lightly. I was *awful* at pre-algebra. One day, we had a pop quiz with 4 questions. Needless to say, I received a 0/4... My mother even feared that she would have problems with me doing math, and she accepted that I was going to struggle with it throughout high school. In seventh grade, I was taking geometry, and I hated it. Whether it was due to being cyberspace, my teacher, or me, I just did not understand. I remember receiving a 2/9 and 3/11 on two of my assignments. Needless to say, my parents were less than thrilled. Going back to school definitely helped me learn. I was able to focus on actually taking in the information without distraction from being at home. This dramatically affected how quickly I picked up information.

It was a massive improvement from my previous performances.

Then, one evening in December 2021, I found a video discussing how to find the area underneath curves; I knew nothing about the topic at the time. I clicked in and was immediately engrossed in it. I immediately wanted to take in the information, process it, and learn from it. So, I went into my mom's office, grabbed myself a stack of printer paper and my favorite pencil, and sat down. I recall very few details from the video itself, other than taking notes on Riemann sums and the definition of an integral. Little did I know, this would be the beginning of a passion—primarily for math, but also for learning in general.

The next day, proudly, I walked into my second-period Algebra 2 class, ready to conquer the world with my newfound knowledge. I walked up to the teacher and said that I had watched a video and taken notes on finding the area under curves. He replied, "Oh, that's calculus stuff." I said, "Yeah, I think it is pretty interesting." That was the end of it. I don't exactly know what I expected from it; however, I was still determined to learn more.

Also, in my endless YouTube surfing, I stumbled upon a Stanford lecture on Cosmology. At the time, I had no clue what "Cosmology" was, other than the video that popped up while I was trying to go to sleep.

Immediately, I took a liking not only to the lecture but also to the professor himself. This man, Leonard Susskind, is one of my greatest inspirations in learning. He is a physics professor at Stanford who has written a series of books that are important later on.

This kick-started my love for physics to the point that I reached out to him and asked what it takes to learn physics (mathematically). I still remember what he replied. He told me that he liked to see my generation "carrying the torch" and that I would need some background in calculus or trigonometry, at the very least, in order to succeed. I told him I was only in Algebra 2, and he told me to go to my local library, borrow books, and self-study them until I was ready to begin learning physics. So that is what I did.

I would spend hours watching videos, taking notes, and eventually reading books. I acquired two of my most important books of all time. The first one is a calculus essentials book by Chris McMullen. I would work through this book religiously, and it taught me more than any person ever could. Additionally, I found one YouTuber in particular that I liked: Blackpenredpen. He made numerous calculus videos. Two of my personal favorites were 100 integrals and 100 derivatives. Both of these videos were well over 6 hours long, and I would watch and take notes on them well into the night (at the expense of not sleeping). This carried on throughout the majority of the first half of 9th grade.

The second book that I found was *The Theoretical Minimum* by Leonard Susskind. This book provides the mathematical reasoning behind the physics that was conducted. Here, I was introduced to Lagrangian Mechanics, something that, at the time, was mind-blowing. Again, I worked through that book religiously, and it has become a staple that I still reference and flip through even today.

At this point in time, my interests have shifted toward following the dopamine rush rather than natural progression. I would do what my brain felt was "right" as opposed to the order in which I should learn things. This would lead me to the many different passion projects I have undertaken, whether they are in physics or math. This is what gave me the drive to finally sit down and write. One day, I don't know if this is a wish or a forewarning; I want to compile all of my writings into one cohesive book that can be read from front to back and contains all the information I have written about, starting with Integral Integrals.

With that being said, let's get into the actual book and solve some differential equations.

Contents

Table of Derivatives

Function $f(x)$	Derivative $f'(x)$
c (constant)	0
x^n	nx^{n-1}
$\sqrt{x}$	$\dfrac{1}{2\sqrt{x}}$
e^x	e^x
a^x	$a^x \ln a$
$\ln x$	$\dfrac{1}{x}$
$\log_a x$	$\dfrac{1}{x \ln a}$
$\sin x$	$\cos x$
$\cos x$	$-\sin x$
$\tan x$	$\sec^2 x$
$\cot x$	$-\csc^2 x$
$\sec x$	$\sec x \tan x$
$\csc x$	$-\csc x \cot x$

Rules	Derivative Formula
Product Rule	$\dfrac{d}{dx}[u(x)v(x)] = u'(x)v(x) + u(x)v'(x)$
Quotient Rule	$\dfrac{d}{dx}\left[\dfrac{u(x)}{v(x)}\right] = \dfrac{u'(x)v(x) - u(x)v'(x)}{[v(x)]^2}$
Chain Rule	$\dfrac{d}{dx}[f(g(x))] = f'(g(x)) \cdot g'(x)$

Table of Integrals

Function $f(x)$	**Integral** $\int f(x)\,dx$
c (constant)	$cx + C$
x^n $(n \neq -1)$	$\frac{x^{n+1}}{n+1} + C$
$\sqrt{x}$	$\frac{2}{3}x^{3/2} + C$
e^x	$e^x + C$
a^x	$\frac{a^x}{\ln a} + C$
$\frac{1}{x}$	$\ln\|x\| + C$
$\frac{1}{x \ln a}$	$\log_a \|x\| + C$
$\sin x$	$-\cos x + C$
$\cos x$	$\sin x + C$
$\tan x$	$-\ln\|\cos x\| + C$
$\cot x$	$\ln\|\sin x\| + C$
$\sec x$	$\ln\|\sec x + \tan x\| + C$
$\csc x$	$-\ln\|\csc x + \cot x\| + C$

Rules	**Integral Formula**
Substitution Rule	$\int f(g(x))g'(x)\,dx = \int f(u)\,du$
Integration by Parts	$\int u\,dv = uv - \int v\,du$

Table of Power Series

Function $f(x)$	Power Series	Interval
$\frac{1}{1-x}$	$\sum_{n=0}^{\infty} x^n$	$\|x\| < 1$
$\frac{1}{1+x}$	$\sum_{n=0}^{\infty} (-1)^n x^n$	$\|x\| < 1$
e^x	$\sum_{n=0}^{\infty} \frac{x^n}{n!}$	$(-\infty, \infty)$
$\sin x$	$\sum_{n=0}^{\infty} (-1)^n \frac{x^{2n+1}}{(2n+1)!}$	$(-\infty, \infty)$
$\cos x$	$\sum_{n=0}^{\infty} (-1)^n \frac{x^{2n}}{(2n)!}$	$(-\infty, \infty)$
$\ln(1+x)$	$\sum_{n=1}^{\infty} (-1)^{n+1} \frac{x^n}{n}$	$-1 < x \leq 1$
$\ln(1-x)$	$-\sum_{n=1}^{\infty} \frac{x^n}{n}$	$\|x\| < 1$
$\arctan x$	$\sum_{n=0}^{\infty} (-1)^n \frac{x^{2n+1}}{2n+1}$	$\|x\| \leq 1$
$\frac{1}{(1-x)^k}$	$\sum_{n=0}^{\infty} \binom{n+k-1}{n} x^n$	$\|x\| < 1$

Table of Laplace Transforms

Function $f(t)$	**Laplace Transform** $\mathcal{L}\{f(t)\} = F(s)$
1	$\frac{1}{s}$
t^n	$\frac{n!}{s^{n+1}}$
e^{at}	$\frac{1}{s-a}$
$\sin(at)$	$\frac{a}{s^2+a^2}$
$\cos(at)$	$\frac{s}{s^2+a^2}$
$\sinh(at)$	$\frac{a}{s^2-a^2}$
$\cosh(at)$	$\frac{s}{s^2-a^2}$
$t^n e^{at}$	$\frac{n!}{(s-a)^{n+1}}$
$\delta(t)$ (Dirac delta)	1
$u_c(t)$ (Heaviside step)	$\frac{e^{-cs}}{s}$
Rules	**Transform Property**
Linearity	$\mathcal{L}\{af(t) + bg(t)\} = aF(s) + bG(s)$
First Shift Theorem	$\mathcal{L}\{e^{at} f(t)\} = F(s - a)$
Second Shift Theorem	$\mathcal{L}\{u_c(t)f(t - c)\} = e^{-cs} F(s)$
Differentiation	$\mathcal{L}\{f'(t)\} = sF(s) - f(0)$
Integration	$\mathcal{L}\left\{\int_0^t f(\tau)\, d\tau\right\} = \frac{F(s)}{s}$

Table of Bessel Functions

Function	Expression or Definition
$J_0(x)$	$\sum_{m=0}^{\infty} \frac{(-1)^m}{(m!)^2} \left(\frac{x}{2}\right)^{2m}$
$J_1(x)$	$\sum_{m=0}^{\infty} \frac{(-1)^m}{m!(m+1)!} \left(\frac{x}{2}\right)^{2m+1}$
$J_n(x)$	$\sum_{m=0}^{\infty} \frac{(-1)^m}{m!\,\Gamma(m+n+1)} \left(\frac{x}{2}\right)^{2m+n}$
$I_0(x)$	$\sum_{m=0}^{\infty} \frac{1}{(m!)^2} \left(\frac{x}{2}\right)^{2m}$
$I_1(x)$	$\sum_{m=0}^{\infty} \frac{1}{m!(m+1)!} \left(\frac{x}{2}\right)^{2m+1}$
$I_n(x)$	$\sum_{m=0}^{\infty} \frac{1}{m!\,\Gamma(m+n+1)} \left(\frac{x}{2}\right)^{2m+n}$
$Y_0(x)$	Bessel function of the second kind
$Y_n(x)$	$Y_n(x) = \frac{J_n(x)\cos(n\pi) - J_{-n}(x)}{\sin(n\pi)}$ for non-integer n

Table of Legendre Polynomials

Polynomial $P_n(x)$	**Expression**
$P_0(x)$	1
$P_1(x)$	x
$P_2(x)$	$\frac{1}{2}(3x^2 - 1)$
$P_3(x)$	$\frac{1}{2}(5x^3 - 3x)$
$P_4(x)$	$\frac{1}{8}(35x^4 - 30x^2 + 3)$
$P_5(x)$	$\frac{1}{8}(63x^5 - 70x^3 + 15x)$
$P_6(x)$	$\frac{1}{16}(231x^6 - 315x^4 + 105x^2 - 5)$
General Formula	$P_n(x) = \frac{1}{2^n n!} \frac{d^n}{dx^n}(x^2 - 1)^n$
Orthogonality	$\int_{-1}^{1} P_m(x) P_n(x)\, dx = \begin{cases} 0 & m \neq n \\ \frac{2}{2n+1} & m = n \end{cases}$

Table of Laguerre Polynomials

Polynomial $L_n(x)$	Expression
$L_0(x)$	1
$L_1(x)$	$-x+1$
$L_2(x)$	$\frac{1}{2}(x^2-4x+2)$
$L_3(x)$	$\frac{1}{6}(-x^3+9x^2-18x+6)$
$L_4(x)$	$\frac{1}{24}(x^4-16x^3+72x^2-96x+24)$
General Formula	$L_n(x)=\frac{e^x}{n!}\frac{d^n}{dx^n}\left(e^{-x}x^n\right)$
Orthogonality	$\int_0^\infty L_m(x)L_n(x)\,e^{-x}\,dx=\begin{cases}0 & m\neq n\\ 1 & m=n\end{cases}$

Table of Hermite Polynomials

Polynomial $H_n(x)$	**Expression**
$H_0(x)$	1
$H_1(x)$	$2x$
$H_2(x)$	$4x^2 - 2$
$H_3(x)$	$8x^3 - 12x$
$H_4(x)$	$16x^4 - 48x^2 + 12$
$H_5(x)$	$32x^5 - 160x^3 + 120x$
$H_6(x)$	$64x^6 - 480x^4 + 720x^2 - 120$
General Formula	$H_n(x) = (-1)^n e^{x^2} \frac{d^n}{dx^n}\left(e^{-x^2}\right)$
Orthogonality	$\int_{-\infty}^{\infty} H_m(x) H_n(x)\, e^{-x^2}\, dx = \begin{cases} 0 & m \neq n \\ 2^n n! \sqrt{\pi} & m = n \end{cases}$

Chapter 1

The History

DIFFERENTIAL EQUATIONS have played a fundamental role in the development of math, science, and engineering throughout human history. A differential equation, in short, is an equation relating the rate of change of a function, or its derivative, to how it changes through space and time. These equations have almost limitless uses, from modeling planetary motion, fluid dynamics, quantum particles, population growth, and even the stock market.

Their history is deeply rooted in the creation of calculus in the 17th century by Isaac Newton and Gottfried Wilhelm Leibniz; however, primitive forms of differential equations can be traced back far before their time.

1.1 Ancient and Classical Influences

The earliest ideas that proposed a differential equation can be traced back to the works of ancient Greek mathematicians. Archimedes, who used the method of exhaustion (infinitesimals used to calculate geometric values) can be accredited with the most differential equation-like non-differential equation we had before the times of Newton and Leibniz. Another great example of Greek influence on differential equations comes from Euclid's *Elements*. Despite studying static shapes and objects, tangent lines and instantaneous motion came up. Additionally, Apollonius' conic sections required knowledge of geometric change along a curve, foreshadowing differential calculus. Their biggest inhibition was their philosophical beliefs that static systems were preferred to dynamic ones. For this reason, research and study into rates of change were limited.

Additionally, early insights from Indian and Islamic mathematicians showed an understanding of rates of change and geometric problems that use tangent lines to solve. Also, they used infinite series to approximate functions and their values. Aryabhata and Bhaskara II, two Indian mathematicians, both used rates of change when analyzing planetary motion. They would use early forms of derivatives by taking finite differences to calculate a rate of change. Aryabhata, specifically, proposed sine tables and interpolated between two values. Though they weren't technically differential equations, their works laid the groundwork for the eventual development of calculus hundreds of years later.

As Europe moved into the Renaissance, Niccolò Tartaglia and Galileo Galilei studied projectile motion, acceleration, and uniform gravity. These developments would eventually push Newton to study motion more in depth. Galileo formed a distance vs. time graph, which illustrated rates of change along curves. This set the foundation for derivatives that Newton and Leibniz would build upon.

1.2 The Birth of Calculus

In the late 17th century, one of the most important mathematical ideas was formalized - calculus. This was primarily done by Isaac Newton (1642-1727) and Gottfried Wilhelm Leibniz (1646-1716).

Figure 1.1: Isaac Newton

Isaac Newton's approach was to develop calculus as a method of "fluxions", so he could study motion. He described variables as quantities that could change over time and derived differential equations that could accurately model physical systems, such as the Earth in orbit with the Sun. Combining with his second law of motion as described in *Principia*

Mathematica,

$$F = ma \tag{1.1}$$

He wrote acceleration as the rate of change of velocity. But what is velocity? Velocity is the rate of change of position. So overall, acceleration is the second derivative of position. He would use the notation of dots over the variables. This indicated that the variable was differentiated with respect to time.

$$F = m\frac{d^2x}{dt^2} \tag{1.2}$$

$$F = m\ddot{x} \tag{1.3}$$

Nowadays, if we have a derivative, we often denote it using prime notation.

$$\frac{dy}{dx} = y' \tag{1.4}$$

$$\frac{d^2y}{dx^2} = y'' \tag{1.5}$$

If you have many consecutive derivatives, you can write them using parentheses.

$$\frac{d^ny}{dx^n} = y^{(n)} \tag{1.6}$$

Leibniz independently developed calculus in the same time frame as Newton. He would introduce many different notations that we still use today, such as

$$\frac{dy}{dx} \tag{1.7}$$

He saw derivatives and integrals from a different perspective. He saw them as algebraic tools and generalized differential equations for applications beyond physics. Through this, he developed the chain rule and popularized the use of infinitesimals to describe change in both physical and theoretical quantities.

Figure 1.2: Gottfried Willhelm Leibniz

1.3 Early Uses of Calculus

Armed with calculus, many 18th-century mathematicians would make groundbreaking discoveries, many of which are formulated using differential equations.

The entire Bernoulli family was famous for their developments in describing physical situations as a differential equation, and coming up with methods to solving these equations. Jacob Bernoulli was the first to write and solve differential equations. He famously solved the Brachistochrone Problem and studied exponential growth.

Figure 1.3: Jacob Bernoulli

His brother, Johann Bernoulli, worked on separable and

homogeneous differential equations.

Figure 1.4: Johann Bernoulli

The Bernoulli Differential Equation is named after Johann.

$$\frac{dy}{dy} + P(x)y = Q(x)y^n \tag{1.8}$$

His work on this non-linear equation showed mathematicians that many equations are actually solvable using traditional methods. They would use these mathematical statements to solve different physical problems, from mechanical systems, to optics, to curves of descent much like the Brachistochrone problem.

Leonhard Euler, one of the most successful mathematicians, shaped the field of differential equations in countless ways.

Figure 1.5: Leonhard Euler

He would come up with a method to solve linear differential equations with constant coefficients. Beyond this, he created the ideas behind exact differential equations, integrating factors, and power series solutions. Numerically, he introduced the Euler Method, the first numerical method to solve differential equations with initial values. He used differential equations to describe the motion of pendulums, springs, orbits. He also popularized the use of the Euler-Lagrange Equation, which set the study of partial differential equations (PDEs) into motion.

1.4 Partial Differential Equations

As science progressed, new problems required different uses of differential equations. Some of these involved multiple variables, calling for the birth of the partial differential equation.

In the 18th century, Jean le Rond d'Alembert derived the wave equation:

$$\frac{\partial^2 u}{\partial t^2} = c^2 \frac{\partial^2 u}{\partial x^2} \tag{1.9}$$

This equation models the vibration of strings perfectly. $u(x, t)$ represents the displacement of the string from its equilibrium position at time t and position x. c is the speed at which the wave propagates.

Figure 1.6: Jean le Rond d'Alembert

Joseph Fourier studied how heat flows from warm to cold. He found that there was a remarkable way to model this physical process mathematically.

Figure 1.7: Joseph Fourier

He came up with

$$\frac{\partial u}{\partial t} = \alpha^2 \frac{\partial^2 u}{\partial x^2} \tag{1.10}$$

To solve this, he introduced Fourier series,

$$f(x) = \sum_{n=1}^{\infty} a_n \sin(\frac{n\pi x}{L}) \tag{1.11}$$

showing that complex functions can be expressed as infinite sums of sine and cosine waves. His work laid the groundwork for signal processing, quantum mechanics, and his series are still used today as a method to solve partial differential equations.

Pierre-Simon Laplace extended the partial differential equations into three dimensions.

Figure 1.8: Pierre-Simon Laplace

He came up with his namesake Laplace equation, which models gravitational and electromagnetic potential fields.

$$\nabla^2 \phi = 0 \tag{1.12}$$

He extended upon the theory behind harmonic functions and Green's functions (which are important in boundary problems).

Joseph-Louis Lagrange restructured how we look at Newtonian mechanics by using the principle of least action.

Figure 1.9: Joseph-Louis Lagrange

This led him to the Euler-Lagrange equation.

$$\frac{d}{dt}(\frac{\partial L}{\partial \dot{q}}) - \frac{\partial L}{\partial q} = 0 \tag{1.13}$$

$L = T - V$ is the Lagrangian of the system (kinetic energy minus potential energy), and $q(t)$ are the coordinates used by the system. This may be in terms of radius and an angle, points on the Cartesian plane, or more abstract coordinate systems, possibly.

An extension of this, pioneered by William Rowan Hamilton, introduced a system of first-order differential equations.

They represent a physical system as it evolves through

Figure 1.10: William Rowan Hamilton

phase space.

$$\dot{q}_i = \frac{\partial H}{\partial p_i} \tag{1.14}$$

$$\dot{p}_i = -\frac{\partial H}{\partial q_i} \tag{1.15}$$

H is the Hamiltonian of the system (total energy = kinetic energy + potential energy)

1.5 Rigor in the 19th-century

Augustin-Louis Cauchy pushed for and developed rigorous definitions to calculus and differential equations.

Figure 1.11: Augustin-Louis Cauchy

He gave definitions for limits, continuity, and convergence of infinite series and sequences. This is the basis for the Real Analysis classes that will take or have already taken. He also did this for complex functions, forming the basis of Complex Analysis classes. He also studied differential equations, defining the existence and uniqueness of solutions and the method of successive approximations.

In 1836, Jacques Sturm and Joseph Louisville introduced

theories behind second-order linear differential operators. These have important uses in the vibration of strings and membranes, eigenvalue problems, orthogonal functions, and Fourier series. They extensively studied linear differential equations of the form

$$\frac{d}{dx}(p(x)\frac{dy}{dx}) + (\lambda w(x) - q(x))y = 0 \tag{1.16}$$

This is known as a Sturm-Liouville Problem. $y(x)$ is the unknown function we are looking to solve for, λ is an eigenvalue, and $p(x)$, $w(x)$, and $q(x)$ are functions with specific properties.

Together, the two proved that the eigenvalues λ_n form a discrete set.

$$\lambda_n = \{\lambda_1, \lambda_2, \lambda_3, ..., \lambda_n\} \tag{1.17}$$

The corresponding eigenfunctions $y_n(x)$ are orthogonal with respect to a weight function $w(x)$. Any suitable function can then be expressed as an infinite series of these eigenfunctions.

Lie groups brought symmetry analysis to differential equations by Sophus Lie. These help to reduce differential equations and/or find exact solutions. They bridge group theory and differential equations, two seemingly independent mathematical fields, which are vital to geometry and physics.

1.6 Classical to Modern Theory

Differential equations became crucial to modern physics starting in the 20th century. Starting with Einstein, who wrote a set of partial differential equations for the curvature of space to mass-energy distribution.

$$R_{\mu\nu} - \frac{1}{2}Rg_{\mu\nu} = \frac{8\pi G}{c^4}T_{\mu\nu} \tag{1.18}$$

This equation governs how space-time bends (left-hand side) as a result of the presence of energy and momentum. Energy is directly related to mass through the mass-energy equivalence principle,

$$E = mc^2 \tag{1.19}$$

for a stationary mass. For a mass in motion,

$$E = \sqrt{p^2c^2 + m^2c^4} \tag{1.20}$$

where p is the momentum of the mass.

Additionally, Maxwell's equations describe electromagnetism,

Figure 1.12: Albert Einstein

and are a set of PDEs in vector calculus. They are technically all named for other people individually, and can be written in integral or differential form.

In the 1920s, Erwin Schrödinger defined quantum mechanics. It relates a wave-function to the energy of a particle.

$$i\hbar\frac{\partial\psi}{\partial t} = \hat{H}\psi \tag{1.21}$$

Figure 1.13: Erwin Schrodinger

On a more mathematical aspect, David Hilbert, Frigyes Riesz, and Stefan Banach abstracted function spaces. These function spaces are crucial to finding exact solutions to PDEs. In the 1940s, Laurent Schwartz introduced distributions, allowing for the definitions of discontinuous functions. This allowed physicists to solve PDEs for signal processing, which generally will output discrete, unique outputs.

1.7 The Rise of Computing

Many differential equations that are used in the real world cannot be solved analytically. Thus, we have to numerically approximate their solutions. Through the course of time, we came up with a number of different methods.

1. Euler's method

2. Runge-Kutta methods

3. Finite element and finite difference methods

After the development of modern computers, the use of numerical methods became widespread. Differential equations could be accurately simulated. Aerodynamics, structural engineering, all rely on numerically solved PDEs. Different software like MATLAB, Mathematica, and certain Python packages became useful for writing code to solve differential equations.

1.8 Applications

Differential equations can be applied to almost any STEM-related field. From mathematics and physics, obviously, to economics, biology, chemistry, ecology, immunology, and even neurology.

Biological differential equations model the growth rate of logistic and exponentially growing populations, SIR models represent how a disease may spread, and Hodgkin-Huxley equations are used to model nerve impulses. Differential equations can be used in pharmacology to determine the half-life of a drug in your body and how long it will take to excrete.

Economic equations can be used to model the growth of an economy, interest rates, and how the stock market may act to random (stochastic) fluctuations.

Edward Lorenz, a meteorologist, opened up the world to chaos theory.

This was present all along, however. Many other physicists generations before Lorenz worked on the three-body problem. Tiny changes in the initial conditions of the systems led to massively varying outcomes. This is what defines chaos. Edward Lorenz applied this idea to meteorology. He simplified atmospheric convection into three nonlinear

Figure 1.14: Edward Lorenz

differential equations.

$$\frac{dx}{dt} = \sigma(y - x) \tag{1.22}$$

$$\frac{dy}{dt} = x(\rho - z) - y \tag{1.23}$$

$$\frac{dz}{dt} = xy - \beta z \tag{1.24}$$

Lorenz dubbed this the butterfly effect because he famously said "a butterfly flapping its wings in Brazil might cause a tornado in Texas."

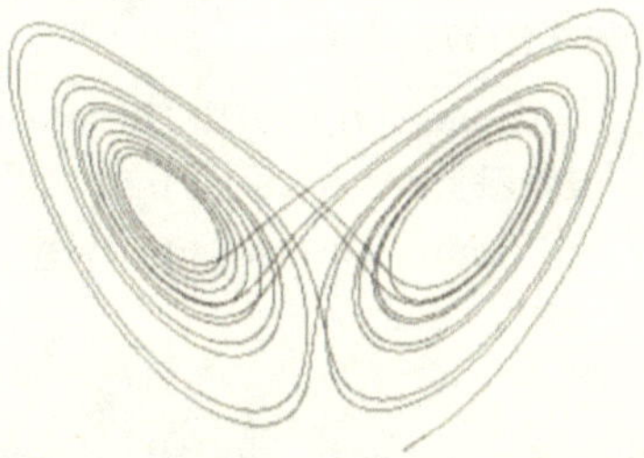

Figure 1.15: Butterfly Effect

Chaos theory further grew and developed into the field of dynamical systems. It studies the shapes or geometries of solutions to differential equations. Phase portraits like Hamilton used show how trajectories evolve over time. The Lorenz attractor is a great example of how changing the initial position and velocity of the point in the attracted field, the outcome vastly differs. It also opened the door to bifurcation theory, which explains why systems change when the parameters vary.

1.9 Afterthought

Differential equations, though mathematical constructs, raise many philosophical questions. Their influence extends far beyond the chalkboard and papers they are written on.

Do differential equations model the deterministic behavior of the universe? Yes. Comparing Newton's laws to the Lorenz attractor, they are completely different in their predictable nature. Newton's laws will give you very similar outputs for similar initial conditions, whereas the Lorenz attractor produces a distinct solution for microscopic changes to the initial conditions. Though we could find out what this solution was, it would be very difficult to predict it given an initial condition. But if we ran the exact same initial conditions ten million times, we would get the exact same output every time. This leads to the idea that differential equations are deterministic but not predictable in nature.

Another problem with this is how much we can divide increments of time and space. Differential equations are based on a continuum, not discrete values like the Planck length, Planck time, etc. that one learns about in physics. So, how does physics, the field based on differential equations, produce statements that contradict the very foundation upon which differential equations were built?

Differential equations have become so ubiquitously known

that they have become a language for people. We use them to describe the physical phenomena experienced by every human, every planet, and every subatomic particle in this universe using differential equations. Galileo once called differential equations "the language in which the universe is written."

Our future is held and relies heavily on differential equations. Though you may not see it at first, every field relies on differential equations in one form or another. They are used for neural networks, artificial intelligence, and quantum computing, which are revolutionizing how computers are created, run, and their processing capabilities. They sit at the heart of every field, and we are learning more and more about them every single day by numerically approximating them and finding analytic solutions. As long as our universe is changing, there will always be a seat for differential equations at the mathematical table.

Chapter 2

Before Solving Them

BEFORE we can actually go about solving a differential equation. there are a few steps that must be covered. These steps will allow us to determine which method we will use to solve a differential equation, how to verify whether a solution is valid, and how to write a general vs. a particular solution given an initial condition. Having this knowledge secured in your mind will greatly benefit you when you are presented with a problem that you need to solve. For example, solving a partial differential equation is vastly different from solving an ordinary differential equation, and solving a linear differential equation is different from solving a homogeneous one. If you don't know what type it is, knowing what steps you need to take to solve it will be ten times more difficult.

2.1 Classifying by Type

Before solving the differential equation, it generally helps to classify it. Certain methods only work on certain types of differential equations, so it is important not to skip this step.

2.1.1 Ordinary vs. Partial

Ordinary Differential Equation

A differential equation is considered ordinary if the derivatives taken are with respect to one variable only. The solution will also be written as a function of one variable.

For example, consider the differential equation

$$\frac{dy}{dx} = 1 + 2\sqrt{x} \tag{2.1}$$

It is ordinary because there is only the derivative of y with respect to x. Do not be mistaken; you can also have multiple derivatives in the same problem. So long as they are all with respect to the same variable, the differential equation is ordinary, such as

$$\frac{d^2y}{dx^2} + 2y\frac{dy}{dx} = \ln(\sin(xy)) \tag{2.2}$$

which is ordinary.

The second type of differential equation is the partial differential equation.

Partial Differential Equation

A differential equation is considered **partial** if the derivatives taken are with respect to multiple variables. The solution will also be written as a function of two or more variables.

An example of this would be the aforementioned wave equation.

$$\frac{\partial^2 u(x,t)}{\partial t^2} = c^2 \frac{\partial^2 u(x,t)}{\partial x^2} \tag{2.3}$$

Make note that the function in question, $u(x,t)$, is a function of both position and time. So, there are partial derivatives with respect to time on the left-hand side and position on the right-hand side.

The function u is not just limited to x and t; however, it encompasses more applications. It can be extended to three dimensions by writing u as $u(x,y,z,t)$, making it a four-variable partial differential equation.

Differentiating between ordinary and partial differential equations is the simplest distinction one can make. Evidently, the notation of the derivatives themselves is different. The use of d indicates a total derivative. The use of ∂ indicates a partial derivative. Looking at which symbol they use will immediately tell you what kind of differential equation you are dealing with.

2.1.2 Order

The next easiest classification to make is the order of the differential equation.

Order

The order of a differential equation is the highest derivative taken in the equation.

An order of one would imply that the highest derivative taken is

$$\frac{dy}{dx} \tag{2.4}$$

An order of two would imply that the highest derivative taken is

$$\frac{d^2y}{dx^2} \tag{2.5}$$

An order of n implies that the highest derivative taken is

$$\frac{d^ny}{dx^n} \tag{2.6}$$

Note: $(\frac{dy}{dx})^n \neq \frac{d^ny}{dx^n}$

The left-hand side implies that it is the first derivative taken to the n-th power. The right-hand side states to take the n-th derivative. These are *very* different, and it is important not to confuse them.

For example, consider

$$\frac{d^2y}{dx^2} - 11x\frac{dy}{dx} + 2y^2 = 0 \tag{2.7}$$

It is a second-order ordinary differential equation.

2.1.3 Linearity

Then, we can classify a differential equation as linear or non-linear.

Linearity

A differential equation is said to be linear if it can be written in the form

$$a_n y^{(n)} + a_{n-1} y^{(n-1)} + ... + a_1 y' + a_0 y = g(x) \quad (2.8)$$

Make note that a_0, a_1, a_2... may be functions of x as well.

So,

$$2xy'' + 9x^4 y' + 4\sqrt{x}y = 0 \quad (2.9)$$

is linear, but

$$2yy' + 3y^2 = 2x \quad (2.10)$$

is non-linear.

This becomes useful because there are certain theorems and techniques that only work for 1st, 2nd, 3rd,... order linear equations. Our hands can often be tied when working with non-linear equations due to their unpredictable and unstable nature.

2.1.4 Homogeneity

Homogeneity is especially important for second and higher order differential equations. There is a trick that we can use for homogeneous differential equations that cannot be used for nonhomogeneous ones.

Homogeneity

A polynomial is homogeneous if and only if the orders of its terms are equal. In other words

$$f(ax, ay) = a^n f(x, y) \tag{2.11}$$

To best illustrate this, we can examine a few polynomials.

$$x^2y + 3y^3 \tag{2.12}$$

Now, you can do this one of two ways. First, you can add up the exponents in each term and compare them, or plug in $x \to ax$ and $y \to ay$ to verify the statement above.

In our example, plugging in ax and ay, we get

$$a^2x^2ay + 3a^3y^3 \tag{2.13}$$

$$= a^3x^2y + 3a^3y^3 \tag{2.14}$$

$$= a^3(x^2y + 3y^3) \tag{2.15}$$

which verifies the statement. Additionally, we could have added the 2 from x^2 and 1 from y^1 and compared them to the 3 from $3y^3$. Since they both have order 3, the polynomial is homogeneous.

Looking at a second example,

$$x + 3xy \tag{2.16}$$

Plugging in, we get

$$a(x + 3axy) \tag{2.17}$$

So, this polynomial is not homogeneous.

Looking at differential equations, a first-order differential equation is homogeneous if

$$f(\frac{y}{x}) = \frac{dy}{dx} \tag{2.18}$$

where $f(\frac{y}{x})$ is a homogeneous function. For example,

$$\frac{y-x}{x-y} = \frac{dy}{dx} \tag{2.19}$$

is homogeneous.

A second-order or higher differential equation is homogeneous if

$$a_n(x)y^{(n)} + ... + a_0(x)y = 0 \tag{2.20}$$

Determining the homogeneity of a differential equation will allow us to make use of characteristic equations, which we will cover further.

2.1.5 Exactness

Exactness generally only applies to first order differential equations. Before we can examine it, though, we need to look at differentials.

Differential Form

The differential form of a differential equation is the original equation written in the form

$$M(x,y)\ dx + N(x,y)\ dy = 0 \tag{2.21}$$

One can convert a differential equation to its exact form by multiplying a first order equation by dx on both sides (technically, it doesn't work this way since $\frac{dy}{dx}$ is not a fraction, but an operator; for our purposes, ignore this). To show how this is done, consider

$$f(x,y)\frac{dy}{dx} + g(x,y) = h(x,y) \tag{2.22}$$

We can multiply dx by both sides to find

$$f(x,y)\ dy + g(x,y)\ dx = h(x,y)\ dx \tag{2.23}$$

Moving the left-hand side over and collecting the terms, we get

$$f(x,y)\ dy\ + (g(x,y) - h(x,y))\ dx = 0 \tag{2.24}$$

We can rearrange these terms, and since they are generic functions, we can change $f(x,y) \to N(x,y)$ and $g(x,y) -$

$h(x, y) \to M(x, y)$. Doing this gives

$$M(x, y)\ dx\ + N(x, y)\ dy = 0 \tag{2.25}$$

So, using this, we can classify a differential equation as exact.

Exactness

A differential equation is said to be exact if given the differential form,

$$M(x, y)\ dx\ + N(x, y)\ dy = 0 \tag{2.26}$$

Then the following relationship holds true for all values of x and y.

$$\partial_y M = \partial_x N \tag{2.27}$$

If a differential equation is exact, then there exists a function $g(x, y)$ such that

$$dg = M\ dx\ + N\ dy \tag{2.28}$$

We will further examine how to use this to our advantage in solving differential equations.

2.1.6 Autonomous

The last type of differential equation we will look at is known as autonomous.

Autonomous

A differential equation is said to be autonomous if the independent variable, x, t does not appear in the differential equation. In other words, the differential equation does not depend on the independent variable.

In cases where the independent variable is time, these differential equations are known as time-invariant differential equations.

The differential equation

$$2\frac{d^2y}{dx^2} - 3\frac{dy}{dx} + 1 = 0 \tag{2.29}$$

is autonomous since x does not explicitly appear anywhere.

2.2 Verifying Solutions

An important part of solving differential equations is verifying that the solution you came up with is actually valid. Much like in algebra, teachers will tell you to plug your value of x back into the original equation to verify that you are correct; the process is very similar. The only difference is that this requires taking derivatives. To demonstrate, consider

$$y'' = -y \tag{2.30}$$

With a solution of $y = \sin x$. To verify this, we can take the second derivative of our proposed solution.

$$y' = \cos x \tag{2.31}$$

$$y'' = -\sin x \tag{2.32}$$

Plugging this into the differential equation, we find

$$-\sin x = -\sin x \tag{2.33}$$

Since the two sides are equal, our solution is valid.

Consider a second example:

$$y = Ce^{-\cos x} \tag{2.34}$$

as a solution to

$$y' = y \sin x \tag{2.35}$$

Taking the derivative of our function, we get

$$y' = C \sin x e^{-\cos x} \tag{2.36}$$

Plugging this in, we get

$$C \sin x e^{-\cos x} = C \sin x e^{-\cos x} \tag{2.37}$$

Make note of how we used C, an arbitrary constant. We could have picked any possible value of C, and it still would have been a solution. Obviously, we can't have multiple solutions to real-life problems, which opens the door to general and particular solutions.

Before we look at the two types of solutions, consider one last example.

$$y = \sqrt{x} \tag{2.38}$$

as a solution to

$$y' - 3 = 2x \tag{2.39}$$

Taking the derivative, we get

$$y' = \frac{1}{2\sqrt{x}} \tag{2.40}$$

Plugging this in, we get

$$\frac{1}{2\sqrt{x}} - 3 = 2x \tag{2.41}$$

Clearly, considering that $x^{-1/2}$ and x are not of the same order, this solution is not valid.

2.3 General vs. Particular Solutions

Considering derivatives, if we have constants involved, their derivatives are simply 0. Additionally, if we have a constant multiplied by a function, then the derivative is simply the derivative of the function times the original constant. This is important because it opens up the door to multiple solutions to a differential equation.

General Solution

The general solution to a differential equation is a solution containing generalized constants C_1, C_2, and so on.

For example, if the solution to the differential equation $y' = 2x$ is

$$y = x^2 \tag{2.42}$$

Then $y = x^2 + 1$, $y = x^2 + 0.1$, and $y = x^2 + 100$ are all solutions. For the general solution, we would write this attached constant as C. Allowing us to rewrite the general solution as

$$y = x^2 + C \tag{2.43}$$

If we have an exponential written in the form

$$y = e^x + C \tag{2.44}$$

Since C is an arbitrary constant, we can rewrite this equation as

$$y = e^x \cdot e^C \tag{2.45}$$

Since this is another constant, we can move it in front of e^x and condense.

$$y = Ce^x \tag{2.46}$$

Particular Solution

The **particular solution** to a differential equation is a solution that passes through a point or has initial conditions $f(c_1) = c_2$.

For a particular solution, we can plug in the x and y values into our general solution and solve for our constants. For a differential equation with n constants, you need n initial conditions in order to solve this.

Apply this to the $y = Ce^x$ case; if we are given the initial condition $y(0) = 1$, then we could plug it in.

$$1 = Ce^0 \tag{2.47}$$

$$C = 1 \tag{2.48}$$

So, the particular solution is

$$y = e^x \tag{2.49}$$

However, if we are given the initial condition $y(1) = 2$, we plug it into the general solution, giving,

$$2 = Ce^1 \tag{2.50}$$

So $C = \frac{2}{e}$
Plugging this back in, we find that

$$y = \frac{2}{e}e^x \tag{2.51}$$

We can rewrite $\frac{2}{e}$ as $2e^{-1}$ and use the exponent rules to simplify this to

$$y = 2e^{x-1} \tag{2.52}$$

2.4 Linearity of Solutions

Many differential equations have multiple distinct solutions. If we have one solution and add another solution to it, then that sum should also be a solution, right? Yes!

> **Linearity of Solutions**
>
> The linear combinations of independent solutions to a differential equation $y_1, y_2, y_3 ..., y_1 + y_2 + y_3 + ...$ are also solutions to the differential equation.

For example, consider

$$y'' = -y \tag{2.53}$$

You can show that $y = \sin x$ and $y = \cos x$ are both solutions. Using these two, we will verify that $y = \sin x + \cos x$ is also a solution.

$$y' = \cos x - \sin x \tag{2.54}$$

$$y'' = -\sin x - \cos x \tag{2.55}$$

Plugging this in, we get the following:

$$-(\sin x + \cos x) = -(\sin x + \cos x) \tag{2.56}$$

This verifies that the linear combination of the two solutions is also a solution to the original differential equation.

2.5 Slope Fields

As a way to describe what a solution to a first-order differential equation looks like, we can use slope fields. Essentially, it allows us to draw vectors pointing in the direction of the derivative at a given point. At each point (x, y), we plug in and find $\frac{dy}{dx}$ and draw a vector pointing in that direction. The higher the slope, the more vertical the line.

Consider the following example:

Example

Draw a slope field for the following differential equation

$$\frac{dy}{dx} = x - y \tag{2.57}$$

So, we can start by inserting integer values, beginning with the origin.

$$(0, 0) \to y' = 0$$
$$(1, 0) \to y' = 1$$
$$(2, 0) \to y' = 2$$
$$(0, 1) \to y' = -1$$
$$(0, 2) \to y' = -2$$

and so on... Then, by drawing vectors based on the magnitude of $\frac{dy}{dx}$, we can construct the slope field. Using a GeoGebra simulation, it generates the following slope field:

Using a slope field, we can approximate what our solution

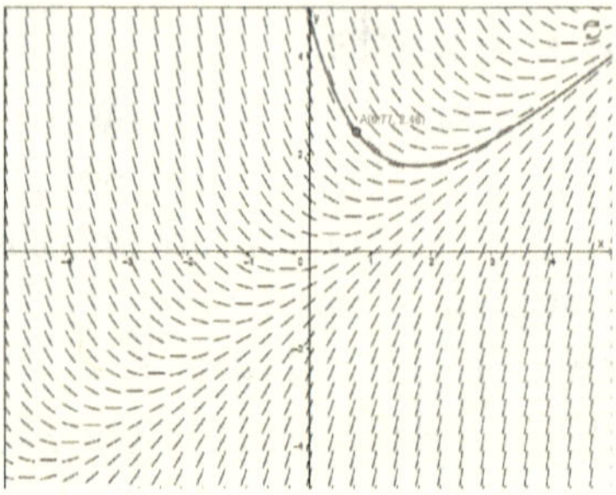

Figure 2.1: Slope field of $\frac{dy}{dx} = x - y$

should look like. If you were to solve a differential equation, plot it, and it is not remotely close to the slope field you can roughly draw, then something most likely went wrong in your work.

2.6 Examples

To wrap up the chapter, we will go through a few examples, practicing the skills learned.

Example 1

Classify the following differential equation:

$$2\frac{dy}{dx} + 2yx = 0 \tag{2.58}$$

Immediately, we can see that, since it is $\frac{dy}{dx}$ as the highest derivative, it is a first-order ordinary differential equation. It is linear since the y and $\frac{dy}{dx}$ terms are all of exponential order 1.

Additionally, we can show that it is homogeneous.

$$\frac{dy}{dx} = -xy \tag{2.59}$$

To write it in $f(\frac{y}{x})$ form, we take it to be

$$\frac{dy}{dx} = \frac{-y}{\frac{1}{x}} \tag{2.60}$$

Due to the 0 on the right-hand side, this is homogeneous. We can convert it to differential form.

$$2\ dy\ + 2xy\ dx = 0 \tag{2.61}$$

$$2xy\ dx\ + 2\ dy = 0 \tag{2.62}$$

Now, we test to see if it is exact.

$$\partial_y M = 2x \tag{2.63}$$

$$\partial N_x = 0 \tag{2.64}$$

So, the differential equation is not exact. Overall, this is an ordinary, first-order, homogeneous linear differential equation.

Example 2

Classify the following differential equation:

$$(\frac{dy}{dx})^3 - \frac{d^2y}{dx^2} = 2\sqrt{x} \tag{2.65}$$

Immediately, we see that this is ordinary. Do not be fooled by the first term; though it may look like this is third order, that term is just the first derivative cubed. So, this is second-order. Since the $\frac{dy}{dx}$ term is raised to a power, this is nonlinear, and since the right-hand side is not equal to 0, it is nonhomogeneous. Overall, it is an ordinary second-order, nonlinear, nonhomogeneous differential equation.

Example 3

Classify the following differential equation:

$$2y'' - 3y' + 5y^2 = 0 \tag{2.66}$$

This differential equation is ordinary, second-order, and nonlinear due to the y^2 term. It is also autonomous, since we see no x in the equation itself. It is also homogeneous since the right-hand side is equal to 0.

Example 4

For the differential equation,

$$y'' - y = 0 \tag{2.67}$$

verify the solution

$$y = C_1 e^x + C_2 e^{-x} \tag{2.68}$$

We can take the second derivative of our proposed solution.

$$y' = C_1 e^x - C_2 e^{-x} \tag{2.69}$$

$$y'' = C_1 e^x + C_2 e^{-x} \tag{2.70}$$

Plugging this back in, we get

$$C_1 e^x + C_2 e^x - (C_1 e^x + C_2 e^{-x}) = 0 \tag{2.71}$$

The terms cancel out, leaving us with

$$0 = 0 \tag{2.72}$$

verifying this solution.

Example 5

For the differential equation,

$$4y^2(y')^2 = 1 \tag{2.73}$$

verify the solution

$$y = \sqrt{x + C} \tag{2.74}$$

Taking the derivative, we get

$$y' = \frac{1}{2\sqrt{x + C}} \tag{2.75}$$

We can square this, allowing us to plug it in.

$$(y')^2 = \frac{1}{4(x + C)} \tag{2.76}$$

We can also find out what $4y^2$ equals in terms of our solution.

$$4y^2 = 4(\sqrt{x + C})^2 \tag{2.77}$$

$$4y^2 = 4(x + C) \tag{2.78}$$

Now we can plug both terms into the differential equation.

$$4(x + C)(\frac{1}{4(x + C)}) = 1 \tag{2.79}$$

The terms cancel out, leaving

$$1 = 1 \tag{2.80}$$

verifying our solution.

Example 4

For the general solution to a differential equation,

$$y = C_1 e^x + C_2 e^{-x} \tag{2.81}$$

and the initial conditions

$$y(0) = 2 \tag{2.82}$$

and

$$y(1) = 3 \tag{2.83}$$

Write the particular solution

The first step we can take is to evaluate the solution at each of the initial conditions. Then, we can write a system of linear equations that we can solve.

$$C_1 e^0 + C_2 e^0 = 2 \tag{2.84}$$

$$C_1 + C_2 = 2 \tag{2.85}$$

And the second equation,

$$C_1 e + \frac{C_2}{e} = 3 \tag{2.86}$$

First, we will solve for C_2. Start with

$$C_1 = 2 - C_2 \tag{2.87}$$

which we can plug into the second equation.

$$3 = 2e - C_2 e + \frac{C_2}{e} \tag{2.88}$$

$$3 - 2e = \frac{C_2(1 - e^2)}{e} \tag{2.89}$$

$$3e - 2e^2 = C_2(1 - e^2) \tag{2.90}$$

So,

$$C_2 = \frac{3e - 2e^2}{1 - e^2} \tag{2.91}$$

Now, we can use the first equation $C_1 + C_2 = 2$ to find C_1. Since,

$$C_1 = 2 - C_2 \tag{2.92}$$

$$C_1 = 2 - \frac{3e - 2e^2}{1 - e^2} \tag{2.93}$$

Last, we will write the particular solution using these constants.

$$y_p = (2 - \frac{3e - 2e^2}{1 - e^2})e^x + \frac{3e - 2e^2}{1 - e^2}e^{-x} \tag{2.94}$$

Chapter 3

First-Order Techniques

FIRST-ORDER DIFFERENTIAL EQUATIONS are the simplest; however, they still have their tricks. Many people think that a first-order differential equation is automatically easier than a second-order, third-order, etc. However, that is not always the case. Some of the most famous first-order differential equations have no known analytic solutions and must be solved numerically.

We, however, can look at some methods that people use to analytically solve the simpler first-order differential equations. These methods were almost all pioneered by Leonhard Euler in the 1700s.

3.1 Separation of Variables

Separation of Variables is a method that we can use to solve first order (and certain higher order) differential equations by breaking apart the equation algebraically.

Separation of Variables

For a differential equation written in the form

$$\frac{dy}{dx} = a(x)b(y) \tag{3.1}$$

We can "separate the variables" by collecting all x terms on one side and all y terms on the other side.

$$\frac{1}{b(y)}\ dy = a(x)\ dx \tag{3.2}$$

This equation also holds true if $y = f(x)$.

Technically, we can't treat $\frac{dy}{dx}$ as a fraction, since it is an operator; however, this statement does hold true due to substitution rules for integrals.

Consider the following example:

$$\frac{1}{y}\frac{dy}{dx} = 2\sin x \tag{3.3}$$

First, we can multiply by dx on both sides since the x terms and the y terms are both independently grouped together.

$$\frac{1}{y}\ dy = 2\sin x\ dx \tag{3.4}$$

Next, integrate on both sides with respect to y on the left and x on the right.

$$\ln y = -2\cos x + C \tag{3.5}$$

Lastly, we can solve for y.

$$y = Ce^{-2\cos x} \tag{3.6}$$

3.2 Linear Differential Equations

We saw linear differential equations in the last chapter when learning how to identify and classify them. Recall that both y and $\frac{dy}{dx}$ terms are linear and are not multiplied by powers of y, nor are they raised to a power themselves. So, a first-order linear differential equation takes the form

$$y' + P(x)y = Q(x) \tag{3.7}$$

To solve this, we use what is known as an integrating factor to simplify this down to a solvable equation. Let $M(x)$ be an integrating factor such that

$$M(x)P(x) = M'(x) \tag{3.8}$$

Now, we multiply the differential equation by this factor.

$$M(x)(y' + P(x)y) \tag{3.9}$$

$$M(x)y' + M'(x)y \tag{3.10}$$

We can determine what exactly $M(x)$ is by solving the original differential equation involving $M(x)$.

$$M(x)P(x) = M'(x) \tag{3.11}$$

$$P(x) = \frac{M'(x)}{M(x)} \tag{3.12}$$

Integrating on both sides gives

$$\int P(x)\, dx = \ln(M(x)) + C \tag{3.13}$$

$$M(x) = Ce^{\int P(x)\, dx} \tag{3.14}$$

Going back to the multiplied version of the differential equation we are looking to solve,

$$M(x)y' + P(x)M(x)y = M(x)Q(x) \tag{3.15}$$

You may notice that the right-hand side is the product rule for $M(x)y$.

$$\frac{d}{dx}(M(x)y) = M(x)y' + M'(x)y \tag{3.16}$$

So, we can rewrite this and integrate it on both sides.

$$\int \frac{d}{dx}(M(x)y) = \int Q(x)M(x) \tag{3.17}$$

$$M(x)y = \int Q(x)Ce^{\int P(x)\, dx}\, dx + C \tag{3.18}$$

Lastly, we divide by $M(x)$ to obtain our final answer.

$$y = Ce^{-\int P(x)} \int Q(x)e^{\int P(x)\, dx}\, dx + Ce^{-\int P(x)} \tag{3.19}$$

3.3 Homogeneous

Consider that we have a differential equation of the form

$$y' + P(x)y = 0 \tag{3.20}$$

We can solve this in one of two ways. First, we can separate it.

$$\frac{dy}{dx} = -P(x)y \tag{3.21}$$

$$\frac{1}{y}\ dy = -P(x)\ dx \tag{3.22}$$

Integrating,

$$\ln y = -\int P(x)\ dx \tag{3.23}$$

$$y = Ce^{-\int P(x)\ dx} \tag{3.24}$$

Our other option was to apply the formula for the linear differential equation, but set $Q(x)$ to be 0. The first integral term cancels out, leaving only

$$y = Ce^{-\int P(x)\ dx} \tag{3.25}$$

3.4 Exact Differential Equations

Recall that an exact differential equation is one of the forms

$$M(x, y)\ dx + N(x, y)\ dy = 0 \tag{3.26}$$

Such that

$$\partial_y M = \partial_x N \tag{3.27}$$

First, take the integral of $M(x, y)$ with respect to x. Since y was treated as a constant, we can write our integration constant as a function of y.

$$F(x, y) = \int M(x, y)\ dx + g(y) \tag{3.28}$$

Next, we can differentiate this with respect to y.

$$\partial_y F = \partial_y(\int M(x, y)\ dx) + g(y)) \tag{3.29}$$

$$\partial_y F = \partial_y(\int M(x, y))\ dx + g'(y) \tag{3.30}$$

We can now equate the right-hand side to $N(x, y)$.

$$\partial_y(\int M(x, y)\ dx) + g'(y) = N(x, y) \tag{3.31}$$

Solve for $g'(y)$ and integrate to find $g(y)$. Last, plug it back into the equation.

$$F(x, y) = \int M(x, y)\ dx + g(y) \tag{3.32}$$

The solution, then, is given by

$$F(x, y) = C \tag{3.33}$$

There were many steps happening, so it is best to walk through it with an actual equation.

$$(2xy + y^3)\ dx\ + (x^2 + 3xy^2)\ dy = 0 \tag{3.34}$$

The first step is always to verify that it is exact.

$$\partial_y M = 2x + 3y^2 \tag{3.35}$$

$$\partial_x N = 2x + 3y^2 \tag{3.36}$$

Since these are equal, it is exact.

So, we write

$$F(x, y) = \int (2xy + y^3)\ dx \tag{3.37}$$

$$F(x, y) = x^2 y + y^3 x + g(y) \tag{3.38}$$

Now, we take the partial derivative with respect to y.

$$\partial_y F = x^2 + 3y^2 x + g'(y) \tag{3.39}$$

Set this equal to $N(x, y)$.

$$x^2 + 3y^2 x + g'(y) = x^2 + 3xy^2 \tag{3.40}$$

The terms cancel out, leaving only

$$g'(y) = 0 \tag{3.41}$$

Integrating gives that

$$g(y) = C_1 \tag{3.42}$$

Now, we have that

$$F(x, y) = x^2y + y^3x + C_1 \tag{3.43}$$

We set this equal to C.

$$x^2y + y^3x + C_1 = C \tag{3.44}$$

Since C and C_1 are arbitrary constants, we can freely combine them, leaving us with the final answer to the differential equation.

$$x^2y + y^3x = C \tag{3.45}$$

If the differential equation is not exact on its own, then we can force it to be exact by introducing an integrating factor. Much like the linear differential equation, it involves multiplying by exponentials of an integral.

$$f(x) = e^{\int \frac{1}{N}(\partial_y M - \partial_x N)\ dx} \tag{3.46}$$

Multiply the entire differential equation by this factor and then repeat the same steps as before.

3.5 Examples

Example 1

Solve the following differential equation

$$\frac{dy}{dx} = y\sqrt{1+x} \tag{3.47}$$

Immediately, we notice that there is only one term that consists of a function of y times a function of x. So, this differential equation is most likely separable.

$$\frac{1}{y}\frac{dy}{dx} = \sqrt{1+x} \tag{3.48}$$

Now, we can multiply dx by both sides.

$$\frac{1}{y}\,dy = \sqrt{1+x}\,dx \tag{3.49}$$

Integrate,

$$\int \frac{1}{y}\,dy = \int \sqrt{1+x}\,dx \tag{3.50}$$

$$\ln y = \frac{2}{3}(x+1)^{\frac{3}{2}} + C \tag{3.51}$$

Lastly, we can solve for y by exponentiating both sides to obtain our general solution.

$$y = Ce^{\frac{2}{3}(x+1)^{\frac{3}{2}}} \tag{3.52}$$

We could have skipped the separation if we had moved the

term on the right-hand side over.

$$\frac{dy}{dx} - y\sqrt{1+x} = 0 \tag{3.53}$$

You may notice two things: this is linear, and this is homogeneous. So, if you recall, a linear homogeneous equation has the form

$$\frac{dy}{dx} + P(x)y = 0 \tag{3.54}$$

This gives it the general solution,

$$y = Ce^{\int -P(x)\ dx} \tag{3.55}$$

Plugging in $P(x) = \sqrt{1+x}$ gives us the exact same solution as we found by separating the differential equation.

$$y = Ce^{\frac{2}{3}(x+1)^{\frac{3}{2}}} \tag{3.56}$$

Example 2

Solve the following differential equation

$$y^n \frac{dy}{dx} = \frac{1}{1+x^2} \tag{3.57}$$

with $y(0) = 1$

This problem is a generalized form of a differential equation with an initial value that we will use to solve for C at the end. This equation, much like the last example, is separable; however, it is not linear. So, we cannot use either of the two methods. Instead, we must separate it.

$$y^n \, dy = \frac{1}{1+x^2} \, dx \tag{3.58}$$

Integrating this on both sides, we get

$$\frac{y^{n+1}}{n+1} = \arctan x + C \tag{3.59}$$

Before we move on, we can plug in our initial condition to find C. As long as we have y and x in our equation, we can still plug in our initial conditions. It does not have to be in the form $y = f(x)$.

$$\frac{1}{n+1} = \arctan(0) + C \tag{3.60}$$

This leaves us with

$$C = \frac{1}{n+1} \tag{3.61}$$

Now, we can plug this back into our solution.

$$\frac{y^{n+1}}{n+1} = \arctan x + \frac{1}{n+1} \tag{3.62}$$

We can either leave it like this or simplify it.

$$y^{n+1} = (n+1)\arctan x + \frac{1}{n+1} \tag{3.63}$$

$$y^{n+1} = \frac{(n+1)^2 \arctan x + 1}{n+1} \tag{3.64}$$

If we really wanted to get it in the $y = f(x)$ form, we could take the $(n+1)$th root to both sides. But how it is written will suffice.

Example 3

Solve the following differential equation

$$y' + \frac{y}{x} = \ln x \tag{3.65}$$

This equation is not separable since there are x terms with y and x terms without y. It is linear, so we can use the formula. First, we have to construct the integrating factor using the formula that we derived in the previous section.

$$M(x) = e^{\int \frac{1}{x}\ dx} \tag{3.66}$$

$$M(x) = e^{\ln x} = x \tag{3.67}$$

Now, we can plug this into the general formula.

$$y = Ce^{-x} \int x \ln x\ dx + Ce^{-x} \tag{3.68}$$

To tackle the integral, we use integration by parts with

$$u = \ln x \qquad dv = x \tag{3.69}$$

$$du = \frac{1}{x} \qquad v = \frac{1}{2}x^2 \tag{3.70}$$

$$= \int x \ln x\ dx = \frac{1}{2}x^2 \ln x - \frac{1}{4}x^2 \tag{3.71}$$

Lastly, we can plug this into our formula to provide us with the general solution.

$$y = C_1 e^{-x} x^2 \ln x - C_2 x^2 e^{-x} + C_3 e^{-x} \tag{3.72}$$

Example 4

Solve the following differential equation

$$y' - 2xy = 0 \tag{3.73}$$

This is homogeneous, and it is separable. Since we have already done two examples that have been separable, we can use the formula for the homogeneous differential equation.

$$y = Ce^{-\int P(x)\ dx} \tag{3.74}$$

Since $P(x) = -2x$, we get that

$$y = Ce^{-\int -2x\ dx} \tag{3.75}$$

$$y = Ce^{x^2} \tag{3.76}$$

Example 5

Solve the following differential equation

$$(2x^2y^2 - x)\ dx\ + (\frac{4}{3}x^3y - 3)\ dy = 0 \tag{3.77}$$

Since it is already in differential form, we can easily test whether it is exact.

$$\partial_y M = 4x^2y \tag{3.78}$$

$$\partial_x N = 4x^2y \tag{3.79}$$

Yes! So, we can follow the steps. First, integrate $M(x, y)$ with respect to x.

$$F(x, y) = \int (2x^2y^2 - x)\ dx + g(y) \tag{3.80}$$

$$F(x, y) = \frac{2}{3}x^3y^2 - \frac{1}{2}x^2 + g(y) \tag{3.81}$$

Take the partial derivative with respect to y and equate it to $N(x, y)$.

$$\frac{4}{3}x^3y - g'(y) = \frac{4}{3}x^3y - 3 \tag{3.82}$$

The $\frac{4}{3}x^3y$ terms cancel out, leaving us with

$$g'(y) = 3 \tag{3.83}$$

So, we integrate this, giving us

$$g(y) = 3y + C_1 \tag{3.84}$$

We can now write the function $F(x, y)$ and equate it to the constant C_2.

$$\frac{2}{3}x^3y^2 - \frac{1}{2}x^2 + 3y + C_1 = C_2 \tag{3.85}$$

The arbitrary constants can be combined into one constant, C, yielding our final solution to the differential equation.

$$\frac{2}{3}x^3y^2 - \frac{1}{2}x^2 + 3y = C \tag{3.86}$$

Chapter 4

Second-Order Techniques

SECOND-ORDER DIFFERENTIAL EQUATIONS is equally as important as first order. They are used to model a number of physical systems, from springs, pendulums, acceleration, electrical circuits, wave motion, and more. This chapter seeks to provide methods for solving certain second-order equations.

4.1 Homogeneous Constant Coefficient

This class of differential equation is the easiest second-order type to solve. It takes the form

$$ay'' + by' + cy = 0 \tag{4.1}$$

Where a, b, c can be any number, real or complex.

We can go about solving this by making a guess at our solution. Since the derivative of exponential functions is also an exponential function, we can try making the guess

$$y = e^{rx} \tag{4.2}$$

$$y' = re^{rx} \tag{4.3}$$

$$y'' = r^2e^{rx} \tag{4.4}$$

Now, we can substitute this in.

$$ar^2e^{rx} + bre^{rx} + ce^{rx} = 0 \tag{4.5}$$

Since $e^{rx} \neq 0$ for any value of x that we can choose, we can divide it out of each term, effectively deleting it. We are left with what is known as the characteristic equation.

Characteristic Equation

For a linear homogeneous differential equation of the form

$$a_ny^{(n)} + a_{n-1}y^{(n-1)} + ... + a_1y' + a_0y = 0 \tag{4.6}$$

The characteristic equation is given by

$$a_nr^n + a_{n-1}r^{n-1} + ... + a_1r + a_0 = 0 \tag{4.7}$$

This opens up three cases for values of r. We can either have distinct real roots, distinct complex roots, or repeated roots. Going back to our second-order case,

$$ar^2 + br + c = 0 \tag{4.8}$$

If $b^2 - 4ac > 0$ (the discriminant), then

$$y = C_1 e^{r_1 x} + C_2 e^{r_2 x} \tag{4.9}$$

Where r_1 and r_2 are the two solutions to the characteristic equation.

If $b^2 - 4ac < 0$, then there would be an imaginary root. So instead of writing it in $r = a \pm bi$ form, we use Euler's Formula to rewrite it as

$$C_1 e^{ax} e^{ibx} + C_2 e^{ax} e^{-ibx} \tag{4.10}$$

$$y = e^{ax}(C_1 \cos(bx) + C_2 \sin(bx)) \tag{4.11}$$

Lastly, if the roots are repeated, that is, $b^2 - 4ac = 0$, then we take the first case, but we add an x to the second term.

$$y = C_1 e^{r_1 x} + C_2 x e^{r_2 x} \tag{4.12}$$

4.2 Homogeneous Variable Coefficients

A second-order homogeneous differential equation is one that takes the form

$$a(x)y'' + b(x)y' + c(x)y = 0 \tag{4.13}$$

To go about solving these, we use a power series. We can either center it around a point, or if there are no initial conditions, we can center it at $x = 0$. So, if we use

$$y = \sum_{n=0}^{\infty} a_n x^n \tag{4.14}$$

Then, we take the first and second derivatives. Note that this can be applied to higher order differential equations and is not limited to only second-order.

$$y' = \sum_{n=1}^{\infty} a_n n x^{n-1} \tag{4.15}$$

$$y'' = \sum_{n=2}^{\infty} a_n n(n-1)x^{n-2} \tag{4.16}$$

We can re-index this so all of the series have x^n terms (this will be important later).

$$y' = \sum_{n=0}^{\infty} a_{n+1}(n+1)x^n \tag{4.17}$$

$$y'' = \sum_{n=0}^{\infty} a_{n+2}(n+2)(n+1)x^n \tag{4.18}$$

We can plug this into the differential equation, combine it into one series and factor out the x^n term. The process would be slightly different depending on what $a(x)$, $b(x)$, $c(x)$ are. This will slightly change how we re-index the series after originally differentiating them. Since it is a case-by-case situation, we will cross this bridge when we do an example.

$$\sum_{n=0}^{\infty}(a_{n+2}(n+2)(n+1)a(x)+a_{n+1}(n+1)b(x)+a_n c(x))x^n = 0 \tag{4.19}$$

Then, you set the constants equal to 0 and solve for a_{n+2}. Plug in different values to find a_n in terms of a_0 and a_1, two arbitrary constants.

4.3 Substitution

If we have a differential equation of the form

$$y'' + p(x)y' = 0 \tag{4.20}$$

We can make the substitution

$$u = y' \tag{4.21}$$

$$u' = y'' \tag{4.22}$$

This allows us to transform the differential equation into

$$u' + p(x)u(x) = 0 \tag{4.23}$$

This first-order differential equation is separable.

$$\frac{du}{dx} = -p(x)u \tag{4.24}$$

Separating this and integrating gives

$$\ln(u) = -\int p(x)\ dx \tag{4.25}$$

$$u = e^{-\int p(x)\ dx} \tag{4.26}$$

Then, since $u = \frac{dy}{dx}$, we have to integrate one more time to get

$$y = \int e^{\int p(x)\ dx} \tag{4.27}$$

4.4 Variation of Parameters

This method works wonders for nonhomogeneous second-order differential equations. When looking at them, your actual solution is made up of two parts; first is the solution to the homogeneous differential equation that we have already looked at how to solve, and the particular solution. In fact, it is just the sum of these two solutions.

$$y = y_h + y_p \tag{4.28}$$

The particular solution is often much more difficult to solve for, but we have a couple of ways that we can solve for it depending on what the function is. If you have a differential equation of the form

$$y'' + q(x)y' + r(x)y = f(x) \tag{4.29}$$

and y_1 and y_2 are solutions to the homogeneous form

$$y'' + q(x)y' + r(x)y = 0 \tag{4.30}$$

then the particular solution is given by

$$y_p = -y_1 \int \frac{y_2 f(x)}{y_1 y_2' - y_2 y_1'}\, dx + y_2 \int \frac{y_1 f(x)}{y_1 y_2' - y_2 y_1'}\, dx \tag{4.31}$$

4.5 Undetermined Coefficients

Given a differential equation

$$y'' + q(x)y' + r(x)y = f(x) \tag{4.32}$$

To find the particular solution, we make a guess at what we think the solution is based on what $f(x)$ is using arbitrary coefficients that we will solve for.

Nonhomogeneous Term	**Form of Particular Solution y_p**
c (constant)	A
x^n	$A_n x^n + A_{n-1}x^{n-1} + \cdots + A_0$
e^{kt}	Ae^{kt}
$x^n e^{kt}$	$(A_n x^n + \cdots + A_0)e^{kt}$
$\sin(bt)$ or $\cos(bt)$	$A\cos(bt) + B\sin(bt)$
$x^n \sin(bt)$ or $x^n \cos(bt)$	$(A_n x^n + \cdots + A_0)\cos(bt) + (B_n x^n + \cdots + B_0)\sin(bt)$

Once we have picked our guess, we can plug it back into the differential equation and solve for the missing coefficients by grouping terms on the left-hand side with those that have the known coefficients on the right-hand side.

4.6 Ansatz

If we are dealing with an ugly differential equation, then we can throw out an ansatz. What is an ansatz? An ansatz is an assumption or guess about the form of an unknown function in order to force a solution. A way of formalizing the substitution made to create a characteristic equation. We used

$$y = e^{rx} \tag{4.33}$$

It does not have to be this function. Many times we use

$$y = ax^r \tag{4.34}$$

when we have polynomials as coefficients or compositions of functions. There may also be cases where we use

$$r \ln x \tag{4.35}$$

We then plug it into the differential equation and solve for r and any associated coefficients.

4.7 Examples

Example 1

Solve the following differential equation.

$$y'' - 4y' - 12y = 2t^3 - t + 3 \tag{4.36}$$

Our first step will be to solve the homogeneous differential equation, since that is half of the answer to our problem. So, we have to solve

$$y'' - 4y - 12y \tag{4.37}$$

We write our characteristic equation using r's.

$$r^2 - 4r - 12 = 0 \tag{4.38}$$

We can easily factor this and solve for our two roots.

$$(r - 6)(r + 2) = 0 \tag{4.39}$$

$$r_1 = 6 \tag{4.40}$$

$$r_2 = -2 \tag{4.41}$$

So, since these are both positive and distinct, the solution to the homogeneous differential equation is

$$y = C_1e^{6x} + C_2e^{-2x} \tag{4.42}$$

Next, we use the method of undetermined coefficients to determine what the particular solution to the differential equation is. Since the function on the right-hand side is a polynomial of degree 3, we make our guess at the particular solution to be

$$y_p = At^3 + Bt^2 + Ct + D \tag{4.43}$$

Taking the first and second derivatives, we get

$$y_p' = 3At^2 + 2Bt \tag{4.44}$$

$$y_p'' = 6At + 2B \tag{4.45}$$

Now, we substitute this into the original differential equation.

$$6At + 2B - 12At^2 - 8Bt - 4C - 12At^3 - 12Bt^2 \tag{4.46}$$
$$-12Ct - 12D = 2t^3 - t + 3$$

Group by the order of the t term.

$$-12At^3 + (-12A - 12B)t^2 + (6A - 8B - 12C)t \tag{4.47}$$
$$+(2B - 4C - 12D) = 2t^3 - t + 3$$

We set the coefficient of the t^3 term on the left-hand side to the coefficient of the t^3 term on the right-hand side. We do this for t^2, t, and the constant terms as well.

$$-12A = 2 \tag{4.48}$$

$$-12A - 12B = 0 \tag{4.49}$$

$$6A - 8B - 12C = -1 \tag{4.50}$$

$$-12D = 3 \tag{4.51}$$

So, we can solve these 4 equations for each of the unknowns.

Doing this gives us the undetermined coefficients that we can plug back into our guess. This will ultimately give us the particular solution to the nonhomogeneous portion of the differential equation.

$$A = -\frac{1}{6} \tag{4.52}$$

$$B = \frac{1}{6} \tag{4.53}$$

$$C = -\frac{1}{9} \tag{4.54}$$

$$D = -\frac{5}{27} \tag{4.55}$$

Lastly, since we have both the homogeneous and particular solutions, we can plug the coefficients in and add the two solutions together.

$$y = y_h + y_p \tag{4.56}$$

$$y(t) = C_1 e^{6x} + C_2 e^{-2x} - \frac{1}{6}t^3 + \frac{1}{6}t^2 - \frac{1}{9}t - \frac{5}{27} \tag{4.57}$$

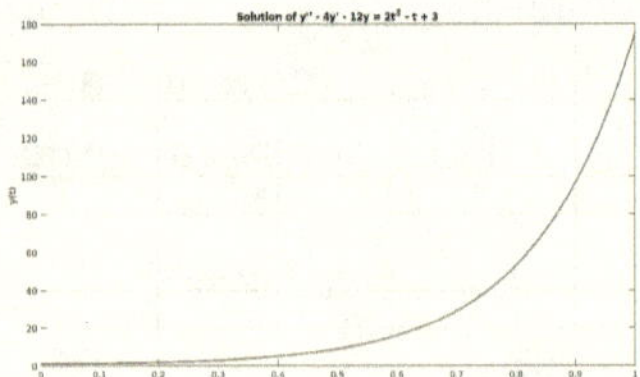

Figure 4.1: Solution to $y'' - 4y' - 12y = 2t^3 - t + 3$

Example 2

Solve the following differential equation.

$$2y'' + 18y = 6\tan 3t \tag{4.58}$$

For this nonhomogeneous differential equation, we can first clean up the constants by dividing both sides by 2.

$$y'' + 9y = 3\tan 3t \tag{4.59}$$

Next, we can solve the homogeneous differential equation.

$$y'' + 9 = 0 \tag{4.60}$$

Writing the characteristic equation,

$$r^2 + 9 = 0 \tag{4.61}$$

In this case, $r_1 = 3i$ and $r_2 = -3i$, so the homogeneous solution is given by

$$y_h = C_1 \cos 3x + C_2 \sin 3x \tag{4.62}$$

Since $3\tan 3t$ isn't a good function for undetermined coefficients, we can use variation of parameters instead. So, we find the Wronskian of our two solutions.

$$W(\cos(3x), \sin(3x)) = \begin{vmatrix} \cos(3x) & \sin(3x) \\ -3\sin(3x) & 3\cos(3x) \end{vmatrix} \tag{4.63}$$

$$= 3\cos^2(3x) + 3\sin^2(3x) = 3 \tag{4.64}$$

So, we write the particular solution y_p.

$$y_p = -\cos 3t \int \frac{3\sin 3t \tan 3t}{3}\, dt + \sin 3t \int \frac{3\cos 3t \tan 3t}{3}\, dt \tag{4.65}$$

Immediately, we can simplify this down to

$$y_p = -\cos 3t \int \frac{1-\cos^2 3t}{\cos^2 3t}\, dt + \sin 3t \int \sin 3t\, dt \tag{4.66}$$

by converting everything into sines and cosines and rewriting $\sin^2 3t$ as $1 - \cos^2 3t$. Then, the second integral is very easy to evaluate.

$$y_p = -\cos 3t \int (\sec 3t - \cos 3t)\, dt + \sin 3t \frac{-\cos 3t}{3} \tag{4.67}$$

Using the integral rules in the table at the beginning of the book, we integrate this. Then, we will have a $\frac{-\cos 3t}{3}$ in both terms that we factor out. This leaves us with

$$y_p = -\frac{\cos 3t}{3}(\ln \| \sec 3t + \tan 3t\| - \sin 3t + \sin 3t) \tag{4.68}$$

The $\sin 3t$'s cancel out, leaving

$$y_p = -\frac{\cos 3t}{3} \ln \| \sec 3t + \tan 3t\| \tag{4.69}$$

Our last step is to combine this with our homogeneous solution y_h. This gives the final answer to this differential equation,

$$y(t) = C_1 \cos 3t + C_2 \sin 3t - \frac{\cos 3t}{3} \ln \| \sec 3t + \tan 3t\| \tag{4.70}$$

Example 3

Solve the following differential equation

$$y'' - 3y = 0 \tag{4.71}$$

with the initial conditions

$$y(0) = 1 \tag{4.72}$$

$$y'(0) = 0 \tag{4.73}$$

This problem is a homogeneous second-order differential equation with an initial value. We can write the characteristic equation.

$$r^2 - 3 = 0 \tag{4.74}$$

Solving this, we find

$$r_1 = \sqrt{3} \tag{4.75}$$

$$r_2 = -\sqrt{3} \tag{4.76}$$

This gives us the general solution,

$$y = C_1 e^{\sqrt{3}x} + C_2 e^{-\sqrt{3}x} \tag{4.77}$$

To find the particular solution based on our initial conditions, we plug in our first condition, $y(0) = 1$.

$$1 = C_1 + C_2 \tag{4.78}$$

Next, we have to differentiate our solution in order to be able to plug in the second condition.

$$y' = C_1\sqrt{3}e^{\sqrt{3}x} - C_2\sqrt{3}e^{-\sqrt{3}x} \tag{4.79}$$

We plug in the condition given $y'(0) = 0$, which gives us

$$0 = C_1\sqrt{3} - C_2\sqrt{3} \tag{4.80}$$

So, this implies that

$$C_1\sqrt{3} = C_2\sqrt{3} \tag{4.81}$$

$$C_1 = C_2 \tag{4.82}$$

And, since $C_1 + C_2 = 1$, then

$$C_1 = C_2 = \frac{1}{2} \tag{4.83}$$

We can now write our particular solution using these constants.

$$y = \frac{1}{2}e^{\sqrt{3}x} + \frac{1}{2}e^{-\sqrt{3}x} \tag{4.84}$$

Example 4

Solve the following differential equation.

$$y'' - xy = 0 \tag{4.85}$$

We begin by assuming a power series solution centered at the origin.

$$y(x) = \sum_{n=0}^{\infty} a_n x^n \tag{4.86}$$

We compute the second derivative term by term.

$$y''(x) = \sum_{n=2}^{\infty} a_n n(n-1)x^{n-2} \tag{4.87}$$

The term xy is given by

$$xy(x) = \sum_{n=0}^{\infty} a_n x^{n+1} = \sum_{n=1}^{\infty} a_{n-1} x^n \tag{4.88}$$

To match powers of x, we reindex the second derivative as follows

$$y''(x) = \sum_{n=0}^{\infty} a_{n+2}(n+2)(n+1)x^n \tag{4.89}$$

Substituting into the original equation, we get

$$\sum_{n=0}^{\infty} \left[a_{n+2}(n+2)(n+1) - a_{n-1}\right] x^n = 0 \tag{4.90}$$

To ensure consistency across indices, define $a_{-1} = 0$

The recurrence relation for the coefficients becomes

$$a_{n+2} = \frac{a_{n-1}}{(n+2)(n+1)} \tag{4.91}$$

This is a three-term recurrence, which means we need to choose three initial values to build the series. But because the equation is second order, we expect only two linearly independent solutions. To ensure this, we set up two basis solutions by choosing specific initial values.

We construct the first solution by setting

$$a_0 = 1, \quad a_1 = 0 \tag{4.92}$$

We use the recurrence to generate all other coefficients.

$$a_2 = \frac{a_{-1}}{2 \cdot 1} = 0 \tag{4.93}$$

$$a_3 = \frac{a_0}{3 \cdot 2} = \frac{1}{6} \tag{4.94}$$

$$a_4 = \frac{a_1}{4 \cdot 3} = 0 \tag{4.95}$$

$$a_5 = \frac{a_2}{5 \cdot 4} = 0 \tag{4.96}$$

$$a_6 = \frac{a_3}{6 \cdot 5} = \frac{1}{180} \tag{4.97}$$

$$a_7 = \frac{a_4}{7 \cdot 6} = 0 \tag{4.98}$$

$$a_8 = \frac{a_5}{8 \cdot 7} = 0 \tag{4.99}$$

$$a_9 = \frac{a_6}{9 \cdot 8} = \frac{1}{12960} \tag{4.100}$$

We now write the first few terms of this solution.

$$y_1(x) = 1 + \frac{x^3}{6} + \frac{x^6}{180} + \frac{x^9}{12960} + \cdots \tag{4.101}$$

This is the power series representation of the Airy function $\mathrm{Ai}(x)$ up to a scaling factor.
We now build the second solution by choosing

$$a_0 = 0, \quad a_1 = 1 \tag{4.102}$$

The recurrence gives

$$a_2 = \frac{a_{-1}}{2 \cdot 1} = 0 \tag{4.103}$$

$$a_3 = \frac{a_0}{3 \cdot 2} = 0 \tag{4.104}$$

$$a_4 = \frac{a_1}{4 \cdot 3} = \frac{1}{12} \tag{4.105}$$

$$a_5 = \frac{a_2}{5 \cdot 4} = 0 \tag{4.106}$$

$$a_6 = \frac{a_3}{6 \cdot 5} = 0 \tag{4.107}$$

$$a_7 = \frac{a_4}{7 \cdot 6} = \frac{1}{504} \tag{4.108}$$

$$a_8 = \frac{a_5}{8 \cdot 7} = 0 \tag{4.109}$$

$$a_9 = \frac{a_6}{9 \cdot 8} = 0 \tag{4.110}$$

$$a_{10} = \frac{a_7}{10 \cdot 9} = \frac{1}{45360} \tag{4.111}$$

We write the second solution

$$y_2(x) = x + \frac{x^4}{12} + \frac{x^7}{504} + \frac{x^{10}}{45360} + \cdots \tag{4.112}$$

This is a second linearly independent solution, corresponding to the Airy function $\mathrm{Bi}(x)$ up to a constant multiple The general solution is a linear combination of the two series.

$$y = C_1 \left(1 + \frac{x^3}{6} + \frac{x^6}{180} + \cdots\right) + C_2 \left(x + \frac{x^4}{12} + \frac{x^7}{504} + \cdots\right) \tag{4.113}$$

Example 5

Solve the following differential equation.

$$y'' - xy' = 0 \tag{4.114}$$

To solve this problem, we can use a substitution.

$$u = y' \quad \Rightarrow \quad y'' = \frac{du}{dx} \tag{4.115}$$

Substituting into the original equation, we get

$$\frac{du}{dx} - xu = 0 \tag{4.116}$$

$$\frac{du}{dx} = xu \tag{4.117}$$

This is now in the form of a separable equation.

$$\frac{1}{u}\frac{du}{dx} = x \tag{4.118}$$

$$\int \frac{1}{u}\, du = \int x\, dx \tag{4.119}$$

$$\ln|u| = \frac{x^2}{2} + C_1 \Rightarrow u = y' = C_2 e^{x^2/2} \tag{4.120}$$

We now integrate $y' = C_2 e^{x^2/2}$ to obtain $y(x)$:

$$y(x) = \int C_2 e^{x^2/2}\, dx + C_3 \tag{4.121}$$

$$y(x) = C_2 \int e^{x^2/2}\, dx + C_3 \tag{4.122}$$

This has no elementary solution, so we must express the integral in terms of special functions. Use the substitution

$$x = \sqrt{2}t \quad \Rightarrow \quad dx = \sqrt{2}\,dt \tag{4.123}$$

$$\int e^{x^2/2}\,dx = \sqrt{2}\int e^{t^2}\,dt \tag{4.124}$$

The imaginary error function is defined as

$$\operatorname{erfi}(x) = \frac{2}{\sqrt{\pi}}\int_0^x e^{t^2}\,dt \tag{4.125}$$

Therefore,

$$\int e^{x^2/2}\,dx = \frac{\sqrt{2\pi}}{2}\operatorname{erfi}\left(\frac{x}{\sqrt{2}}\right) + C \tag{4.126}$$

Using this, we arrive at the final solution.

$$y(x) = C_1 \operatorname{erfi}\left(\frac{x}{\sqrt{2}}\right) + C_2 \tag{4.127}$$

Example 6

Solve the following differential equation.

$$2y'' + 3y' + y = 3\sin t + t^2 \tag{4.128}$$

First, we solve the homogeneous equation,

$$2y'' + 3y' + y = 0 \tag{4.129}$$

The characteristic equation is

$$2r^2 + 3r + 1 = 0 \tag{4.130}$$

$$(2r + 1)(r + 1) = 0 \tag{4.131}$$

$$r = -1, \quad r = -\frac{1}{2} \tag{4.132}$$

So,

$$y_h = C_1 e^{-x} + C_2 e^{-\frac{1}{2}x} \tag{4.133}$$

Since the non-homogeneous term $3\sin t + t^2$ is linear, we can solve for both particular solutions and add them together. Let

$$y_p = At^2 + Bt + C \tag{4.134}$$

$$y_p' = 2At + B \tag{4.135}$$

$$y_p'' = 2A \tag{4.136}$$

Substitute this into the ODE.

$$4A + 6At + 3B + At^2 + Bt + C = t^2 \tag{4.137}$$

Factor each term by its power of t.

$$At^2 + (6A + B)t + (4A + 3B + C) = t^2 \tag{4.138}$$

$$\Rightarrow A = 1, \quad B = 6, \quad C = 14 \tag{4.139}$$

Thus,

$$y_p = t^2 - 6t + 14 \tag{4.140}$$

Next, we will solve for the second part of the particular solution. Make the guess

$$y_p = A\sin t + B\cos t \tag{4.141}$$

$$y_p' = A\cos t - B\sin t \tag{4.142}$$

$$y_p'' = -A\sin t - B\cos t \tag{4.143}$$

Substituting this in, we get that

$$2A\sin t + 2B\cos t + 3A\cos t - 3B\sin t - A\sin t - B\cos t = 3\sin t \tag{4.144}$$

$$(A - 3B)\sin t + (B + 3A)\cos t = 3\sin t \tag{4.145}$$

$$\Rightarrow A - 3B = 3 \qquad B + 3A = 0 \tag{4.146}$$

Solving for the coefficients, we get that

$$A = \frac{3}{10} \qquad B = -\frac{9}{10} \tag{4.147}$$

Thus, the second particular solution is

$$y_p = \frac{3}{10}\sin t - \frac{9}{10}\cos t \tag{4.148}$$

Now that we have the two particular solutions, we combine

them with our homogeneous solution.

$$y = C_1 e^{-x} + C_2 e^{-\frac{1}{2}x} + t^2 - 6t + 14 + \frac{3}{10}\sin t - \frac{9}{10}\cos t \tag{4.149}$$

Example 7

Solve the following differential equation.

$$y'' = x + \sin x \tag{4.150}$$

Since we only have 1 derivative term, we do not have to make any substitutions or work with this. We can simply integrate both sides with respect to x to obtain the first derivative y'.

$$\int y''\,dx = \int (x + \sin x)\,dx \tag{4.151}$$

Evaluating both sides gives

$$y' = \frac{x^2}{2} - \cos x + C_1 \tag{4.152}$$

Now we integrate again to find y.

$$\int y'\,dx = \int \left(\frac{x^2}{2} - \cos x + C_1\right) dx \tag{4.153}$$

We integrate term by term, giving us the general solution to the differential equation.

$$y = \frac{x^3}{6} - \sin x + C_1 x + C_2 \tag{4.154}$$

Example 8

Solve the following differential equation.

$$y'' + 2y(y')^3 = 0 \tag{4.155}$$

We start by making the substitution $u = y' = \frac{dy}{dx}$.

$$u = y' = \frac{dy}{dx} \tag{4.156}$$

Using the chain rule, the second derivative y'' can be expressed as

$$y'' = \frac{du}{dx} = \frac{du}{dy} \cdot \frac{dy}{dx} = u\frac{du}{dy} \tag{4.157}$$

Next, substitute these into the original differential equation.

$$u\frac{du}{dy} + 2yu^3 = 0 \tag{4.158}$$

Assuming $u \neq 0$, divide the equation by u^2 to simplify it.

$$\frac{du}{dy} + 2yu^2 = 0 \tag{4.159}$$

Moving the second term to the right-hand side

$$\frac{du}{dy} = -2yu^2 \tag{4.160}$$

This equation is now separable, and can be easily solved.

$$\frac{du}{u^2} = -2y\,dy \tag{4.161}$$

Integrating both sides gives

$$\int \frac{1}{u^2}\,du = \int -2y\,dy \tag{4.162}$$

Evaluating the integrals results in

$$-\frac{1}{u} = -y^2 + C_1 \tag{4.163}$$

Rearranging the equation to solve for u leads to

$$\frac{1}{u} = y^2 - C_1 \tag{4.164}$$

Recall the definition $u = \frac{dy}{dx}$, so we have

$$\frac{dy}{dx} = \frac{1}{y^2 - C_1} \tag{4.165}$$

Separate variables again to integrate with respect to x

$$(y^2 - C_1)\,dy = dx \tag{4.166}$$

Integrating both sides yields

$$\int (y^2 - C_1)\,dy = \int dx \tag{4.167}$$

Evaluating the integrals gives the general solution.

$$\frac{y^3}{3} - C_1 y = x + C_2 \tag{4.168}$$

Thus, the implicit solution to the original differential equation is

$$\frac{y^3}{3} - C_1 y = x + C_2 \tag{4.169}$$

Example 9

Solve the following differential equation.

$$y'' - 4y' + 3 = \ln t \tag{4.170}$$

$$y'' - 4y' + 3y = \ln t \tag{4.171}$$

We begin with the homogeneous solution. The characteristic equation is:

$$r^2 - 4r + 3 = 0 \tag{4.172}$$

Factoring gives:

$$(r-1)(r-3) = 0 \Rightarrow r = 1, 3 \tag{4.173}$$

So the homogeneous solution is:

$$y_h(t) = C_1 e^t + C_2 e^{3t} \tag{4.174}$$

We now use the variation of parameters. Let the fundamental solutions be:

$$y_1(t) = e^t, \quad y_2(t) = e^{3t} \tag{4.175}$$

Compute the Wronskian:

$$W(t) = \begin{vmatrix} e^t & e^{3t} \\ e^t & 3e^{3t} \end{vmatrix} = e^t \cdot 3e^{3t} - e^t \cdot e^{3t} = 2e^{4t} \tag{4.176}$$

Using the variation of parameters formula:

$$y_p = -y_1(t)\int \frac{y_2(t)\cdot \ln t}{W(t)}dt + y_2(t)\int \frac{y_1(t)\cdot \ln t}{W(t)}dt \quad (4.177)$$

Substitute the values:

$$y_p = -e^t \int \frac{e^{3t}\ln t}{2e^{4t}}dt + e^{3t}\int \frac{e^t \ln t}{2e^{4t}}dt \quad (4.178)$$

Simplify each integrand:

$$y_p = -\frac{1}{2}e^t \int e^{-t}\ln t\,dt + \frac{1}{2}e^{3t}\int e^{-3t}\ln t\,dt \quad (4.179)$$

Now we handle each integral using integration by parts and express them in terms of Ei. Start with:

$$I_1 = \int e^{-t}\ln t\,dt \quad (4.180)$$

Let:

$$u = \ln t, \quad dv = e^{-t}dt \Rightarrow du = \frac{1}{t}dt, \quad v = -e^{-t} \quad (4.181)$$

Then:

$$I_1 = -e^{-t}\ln t + \int \frac{e^{-t}}{t}dt = -e^{-t}\ln t + \mathrm{Ei}(-t) \quad (4.182)$$

Similarly, for:

$$I_2 = \int e^{-3t} \ln t \, dt \tag{4.183}$$

Let:

$$u = \ln t, \quad dv = e^{-3t} dt \Rightarrow du = \frac{1}{t} dt, \quad v = -\frac{1}{3} e^{-3t} \tag{4.184}$$

Then:

$$I_2 = -\frac{1}{3} e^{-3t} \ln t + \frac{1}{3} \int \frac{e^{-3t}}{t} dt = -\frac{1}{3} e^{-3t} \ln t + \frac{1}{3} \operatorname{Ei}(-3t) \tag{4.185}$$

Now plug these into the expression for y_p and simplify.

$$y_p = \frac{1}{2} \ln t + \frac{1}{2} e^t \operatorname{Ei}(-t) - \frac{1}{6} \ln t + \frac{1}{6} e^{3t} \operatorname{Ei}(-3t) \tag{4.186}$$

Combine the $\ln t$ terms.

$$\frac{1}{2} \ln t - \frac{1}{6} \ln t = \frac{1}{3} \ln t \tag{4.187}$$

So, the particular solution becomes

$$y_p = \frac{1}{3} \ln t + \frac{1}{2} e^t \operatorname{Ei}(-t) + \frac{1}{6} e^{3t} \operatorname{Ei}(-3t) \tag{4.188}$$

And the general solution is

$$y(t) = C_1 e^t + C_2 e^{3t} + \frac{1}{3} \ln t + \frac{1}{2} e^t \operatorname{Ei}(-t) + \frac{1}{6} e^{3t} \operatorname{Ei}(-3t) \tag{4.189}$$

Example 10

Solve the following differential equation.

$$x^2y'' + 4xy' + 2y = 3e^{-2x}\sin 2x \tag{4.190}$$

$$x^2y'' + 4xy' + 2y = 3e^{-2x}\sin(2x) \tag{4.191}$$

We begin by recognizing that the entire left-hand side matches the expression for the second derivative of the product x^2y

$$\frac{d^2}{dx^2}(x^2y) = x^2y'' + 4xy' + 2y \tag{4.192}$$

This observation allows us to simplify the original equation by introducing a substitution.

$$z = x^2y \tag{4.193}$$

Using this substitution, the differential equation becomes much simpler.

$$z'' = 3e^{-2x}\sin(2x) \tag{4.194}$$

To solve for z, we integrate both sides once

$$z' = \int 3e^{-2x}\sin(2x)\,dx + C_1 \tag{4.195}$$

The integral on the right is a well-known type. We recall that for constants a and b, the integral of $e^{ax}\sin(bx)$ takes the form

$$\int e^{ax}\sin(bx)\,dx = \frac{e^{ax}}{a^2+b^2}(a\sin bx - b\cos bx) \qquad (4.196)$$

Here $a = -2$ and $b = 2$, so substituting gives

$$\int e^{-2x}\sin(2x)\,dx = \frac{e^{-2x}}{8}(-2\sin 2x - 2\cos 2x) \qquad (4.197)$$

This simplifies to

$$\int e^{-2x}\sin(2x)\,dx = \frac{e^{-2x}}{4}(-\sin 2x - \cos 2x) \qquad (4.198)$$

Multiplying by 3 gives the full expression for z'

$$z' = \frac{3e^{-2x}}{4}(-\sin 2x - \cos 2x) + C_1 \qquad (4.199)$$

Now we integrate again to obtain z

$$z = \int \left[\frac{3e^{-2x}}{4}(-\sin 2x - \cos 2x) + C_1\right] dx + C_2 \qquad (4.200)$$

We split this integral into two parts for clarity.

$$z = \frac{3}{4}\int e^{-2x}(-\sin 2x - \cos 2x)\,dx + C_1x + C_2 \qquad (4.201)$$

We already know the integral of $e^{-2x}\sin(2x)$, so now we compute the integral of $e^{-2x}\cos(2x)$ using the same identity.

$$\int e^{-2x}\cos(2x)\,dx = \frac{e^{-2x}}{8}(-2\cos 2x + 2\sin 2x) \tag{4.202}$$

This simplifies to

$$\int e^{-2x}\cos(2x)\,dx = \frac{e^{-2x}}{4}(\sin 2x - \cos 2x) \tag{4.203}$$

Putting everything together, we obtain

$$z = \frac{3e^{-2x}}{16}(\cos 2x - \sin 2x) + C_1 x + C_2 \tag{4.204}$$

We recall that our substitution was $z = x^2 y$, so now we return to the original variable.

$$x^2 y = \frac{3e^{-2x}}{16}(\cos 2x - \sin 2x) + C_1 x + C_2 \tag{4.205}$$

Finally, we divide both sides by x^2 to isolate the function $y(x)$

$$y(x) = \frac{3e^{-2x}}{16x^2}(\cos 2x - \sin 2x) + \frac{C_1}{x} + \frac{C_2}{x^2} \tag{4.206}$$

Example 11

Solve the following differential equation.

$$f'' = f^{-1}(x) \tag{4.207}$$

$$f''(x) = f^{-1}(x) \tag{4.208}$$

Assume a power function solution

$$f(x) = ax^r \tag{4.209}$$

Compute the second derivative

$$f''(x) = ar(r-1)x^{r-2} \tag{4.210}$$

Express the functional inverse f^{-1}:

$$y = ax^r \implies x = \left(\frac{y}{a}\right)^{\frac{1}{r}} \implies f^{-1}(x) = \left(\frac{x}{a}\right)^{\frac{1}{r}} \tag{4.211}$$

Substitute into the differential equation

$$ar(r-1)x^{r-2} = a^{-\frac{1}{r}}x^{\frac{1}{r}} \tag{4.212}$$

For these to match, the powers of x must be the same, so we set them equal and solve for r.

$$r - 2 = \frac{1}{r} \implies r^2 - 2r = 1 \implies r^2 - 2r - 1 = 0 \tag{4.213}$$

We can solve this using the quadratic formula.

$$r = \frac{2 \pm \sqrt{4+4}}{2} = 1 \pm \sqrt{2} \tag{4.214}$$

Now, we have to match the coefficients on both sides to solve for a.

$$a^{1+\frac{1}{r}} = r(r-1) \tag{4.215}$$

Solve for a.

$$a = (r(r-1))^{\frac{r}{r+1}} \tag{4.216}$$

For the root

$$r = 1 + \sqrt{2} \tag{4.217}$$

calculate denominator.

$$r + 1 = 2 + \sqrt{2} \tag{4.218}$$

Rationalize the exponent

$$\frac{r}{r+1} = \frac{1+\sqrt{2}}{2+\sqrt{2}} \times \frac{2-\sqrt{2}}{2-\sqrt{2}} = \frac{\sqrt{2}}{2} = \frac{1}{\sqrt{2}} \tag{4.219}$$

Calculate base

$$r(r-1) = (1+\sqrt{2}) \times \sqrt{2} = \sqrt{2} + 2 \tag{4.220}$$

Therefore,

$$a = \left(\sqrt{2}+2\right)^{\frac{1}{\sqrt{2}}} \tag{4.221}$$

For the root

$$r = 1 - \sqrt{2} \tag{4.222}$$

calculate denominator

$$r + 1 = 2 - \sqrt{2} \tag{4.223}$$

Rationalize the exponent

$$\frac{r}{r+1} = \frac{1-\sqrt{2}}{2-\sqrt{2}} \times \frac{2+\sqrt{2}}{2+\sqrt{2}} = \frac{-\sqrt{2}}{2} = -\frac{1}{\sqrt{2}} \tag{4.224}$$

Calculate base

$$r(r-1) = (1-\sqrt{2})(-\sqrt{2}) = -\sqrt{2} + 2 = 2 - \sqrt{2} \tag{4.225}$$

Therefore,

$$a = \left(2 - \sqrt{2}\right)^{-\frac{1}{\sqrt{2}}} \tag{4.226}$$

$$f(x) = \left(\sqrt{2} + 2\right)^{\frac{1}{\sqrt{2}}} x^{1+\sqrt{2}} \tag{4.227}$$

$$f(x) = \left(2 - \sqrt{2}\right)^{-\frac{1}{\sqrt{2}}} x^{1-\sqrt{2}} \tag{4.228}$$

Example 12

Solve the following differential equation.

$$y'' = \frac{1}{y'(y(x))} \tag{4.229}$$

$$y''(x) = \frac{1}{y'\left(y(x)\right)} \tag{4.230}$$

Assume a power function form for the solution

$$y(x) = ax^r \tag{4.231}$$

Compute the first and second derivatives.

$$y'(x) = arx^{r-1}, \quad y''(x) = ar(r-1)x^{r-2} \tag{4.232}$$

Evaluate the composition inside the right-hand side

$$y(x) = ax^r \quad \Rightarrow \quad y'(y(x)) = ar\left(ax^r\right)^{r-1} = ara^{r-1}x^{r(r-1)} \tag{4.233}$$

Take the reciprocal to match the differential equation

$$\frac{1}{y'(y(x))} = \frac{1}{a^r r x^{r(r-1)}} \tag{4.234}$$

Substitute both sides into the original equation

$$ar(r-1)x^{r-2} = \frac{1}{a^r r x^{r(r-1)}} \tag{4.235}$$

Multiply both sides by $a^r r x^{r(r-1)}$ to eliminate the denominator

$$a^{r+1}r^2(r-1)x^{r-2+r(r-1)} = 1 \tag{4.236}$$

Simplify the exponent

$$r-2+r(r-1) = r-2+r^2-r = r^2-2 \tag{4.237}$$

So the equation becomes

$$a^{r+1}r^2(r-1)x^{r^2-2} = 1 \tag{4.238}$$

For this to be valid for all x, the exponent must be zero

$$r^2-2=0 \quad \Rightarrow \quad r=\pm\sqrt{2} \tag{4.239}$$

Now solve for a by plugging $r=\sqrt{2}$ and $r=-\sqrt{2}$ into

$$a^{r+1}r^2(r-1) = 1 \tag{4.240}$$

For $r=\sqrt{2}$:

$$a^{\sqrt{2}+1}(\sqrt{2})^2(\sqrt{2}-1) = 1 \tag{4.241}$$

$$a^{\sqrt{2}+1}\cdot 2(\sqrt{2}-1) = 1 \quad \Rightarrow \quad a^{\sqrt{2}+1} = \frac{1}{2(\sqrt{2}-1)} \tag{4.242}$$

$$a = \left(\frac{1}{2(\sqrt{2}-1)}\right)^{\frac{1}{\sqrt{2}+1}} \tag{4.243}$$

Rationalize the base

$$2(\sqrt{2}-1) = \frac{2(\sqrt{2}-1)(\sqrt{2}+1)}{\sqrt{2}+1} = \frac{2(2-1)}{\sqrt{2}+1} = \frac{2}{\sqrt{2}+1} \tag{4.244}$$

So

$$a = \left(\frac{1}{\frac{2}{\sqrt{2}+1}}\right)^{\frac{1}{\sqrt{2}+1}} = \left(\frac{\sqrt{2}+1}{2}\right)^{\frac{1}{\sqrt{2}+1}} \tag{4.245}$$

Similarly, for $r = -\sqrt{2}$:

$$a^{-\sqrt{2}+1}(\sqrt{2})^2(-\sqrt{2}-1) = 1 \tag{4.246}$$

$$a^{1-\sqrt{2}} \cdot 2(-\sqrt{2}-1) = 1 \quad \Rightarrow \quad a^{1-\sqrt{2}} = \frac{-1}{2(\sqrt{2}+1)} \tag{4.247}$$

This gives a negative base, which could still be valid in a generalized or complex sense but isn't real-valued for all x, so we discard it for real-valued solutions.
Final real-valued solution

$$y(x) = \left(\frac{\sqrt{2}+1}{2}\right)^{\frac{1}{\sqrt{2}+1}} x^{\sqrt{2}} \tag{4.248}$$

Chapter 5

Higher-Order Techniques

HIGHER-ORDER DIFFERENTIAL EQUATIONS are limited in their solution capabilities due to their instability. Generally, especially for non-linear ones, they are unsolvable by traditional techniques. Because of this, we usually have to rely on numerical methods to approximate solutions. There are some instances where we can use the second order techniques, such as characteristic equations, undetermined coefficients, and variation of parameters to solve them. We can also make guesses as we will see. The majority of the solvable differential equations of higher order are of the form

$$a_n y^{(n)} + a_{n-1} y^{(n-1)} + ... + a_o y = f(x) \tag{5.1}$$

5.1 Convert to a System

The first tactic we can use is converting to a system of first-order equations. Consider the following fourth-order differential equation.

$$x'''' - 7x''' + 4x'' + 5x' - 2x = 0 \tag{5.2}$$

To use this method, we make a series of substitutions.

$$x_1 = x' \tag{5.3}$$

$$x_2 = x_1' = x'' \tag{5.4}$$

$$x_3 = x_2' = x'' \tag{5.5}$$

$$x_4 = x_3' = x''' \tag{5.6}$$

$$x_4' = x_3' = x'''' \tag{5.7}$$

Now, we substitute this in and solve for x_4'.

$$x_4' = 2x_1 - 5x_2 - 4x_3 + 7x_4 \tag{5.8}$$

By differentiating each of the x_n's, we get

$$x_1' = 0x_1 + x_2 + 0x_3 + 0x_4 \tag{5.9}$$

$$x_2' = 0x_1 + 0x_2 + x_3 + 0x_4 \tag{5.10}$$

$$0x_1 + 0x_2 + 0x_3 + x_4 \tag{5.11}$$

$$x_4' = 2x_1 - 5x_2 - 4x_3 + 7x_4 \tag{5.12}$$

We can write these entries in a matrix.

$$\mathbf{x} = \begin{bmatrix} x_1 \\ x_2 \\ x_3 \\ x_4 \end{bmatrix} = \begin{bmatrix} x \\ x' \\ x'' \\ x''' \end{bmatrix} \tag{5.13}$$

Then the system becomes:

$$\frac{d\mathbf{x}}{dt} = \begin{bmatrix} 0 & 1 & 0 & 0 \\ 0 & 0 & 1 & 0 \\ 0 & 0 & 0 & 1 \\ 2 & -5 & -4 & 7 \end{bmatrix} \mathbf{x}$$

Let A denote the coefficient matrix. To analyze the system, we compute the eigenvalues of A by solving:

$$\det(A - \lambda I) = 0$$

We compute:

$$\det \left(\begin{bmatrix} -\lambda & 1 & 0 & 0 \\ 0 & -\lambda & 1 & 0 \\ 0 & 0 & -\lambda & 1 \\ 2 & -5 & -4 & 7-\lambda \end{bmatrix} \right)$$

Use cofactor expansion or a determinant formula for a companion matrix. This determinant yields the characteristic polynomial:

$$p(\lambda) = \lambda^4 - 7\lambda^3 + 4\lambda^2 + 5\lambda - 2$$

We solve:

$$\lambda^4 - 7\lambda^3 + 4\lambda^2 + 5\lambda - 2 = 0$$

You may notice that this is the exact same equation we

would be solving if we had used characteristic equations instead. You would be right. This is not factorable, so you would have to use numerical methods to approximate the roots. Assuming the eigenvalues are distinct real or complex roots.

$$\lambda_1, \lambda_2, \lambda_3, \lambda_4$$

The general solution is

$$\mathbf{x}(t) = c_1\mathbf{v}_1e^{\lambda_1 t} + c_2\mathbf{v}_2e^{\lambda_2 t} + c_3\mathbf{v}_3e^{\lambda_3 t} + c_4\mathbf{v}_4e^{\lambda_4 t}$$

where $\mathbf{v}_i$ are the corresponding eigenvectors of matrix A, and c_i are the general constants.

As an extension of eigenvalues, we can sketch what are known as phase portraits for systems of differential equations. For example, if we have a second-order differential equation written in the form

$$x' = y \tag{5.14}$$

$$y' = -x \tag{5.15}$$

In order to sketch our diagram, we write this system as a matrix and find the eigenvalues. In this case, they are at $\pm i$. Considering the table below, we would have a circular orbit around the origin.

Eigenvalues	**Trace/Determinant**	**Phase Portrait**
λ_1, λ_2 real, distinct, both < 0	tr < 0, det > 0	Inward straight lines
λ_1, λ_2 real, distinct, both > 0	tr > 0, det > 0	Outward straight lines
$\lambda_1 < 0 < \lambda_2$ (real)	det < 0	To and from eig. directions
$\lambda = a \pm bi$, $a < 0$	tr < 0, det > 0	Inward spirals
$\lambda = a \pm bi$, $a > 0$	tr > 0, det > 0	Outward spirals
$\lambda = \pm bi$	tr $= 0$, det > 0	Closed orbits
Real, repeated $\lambda < 0$	tr < 0	Curved inward paths
Real, repeated $\lambda > 0$	tr > 0	Curved outward paths

5.2 Laplace Transform

The Laplace Transform converts a function of time $f(t)$ into a function of a complex variable s.

$$\mathcal{L}\{f(t)\} = F(s) = \int_0^\infty e^{-st} f(t)\, dt$$

- $f(t)$ is a time-domain function.
- $F(s)$ is the Laplace-domain version of $f(t)$.

By using a Laplace Transform, it simplifies our initial value problem by transforming a differential equation into an algebraic equation. They are useful for linear differential equations with constant coefficients, handling all initial value problems, and modeling discontinuous inputs. They have countless uses in circuit analysis, harmonics, and other fields in physics.

For differential equations, especially, the following is helpful.

$$\mathcal{L}\{f'(t)\} = sF(s) - f(0), \quad \mathcal{L}\{f''(t)\} = s^2F(s) - sf(0) - f'(0)$$

Example

Solve the IVP

$$y'' + 3y' + 2y = 0, \qquad y(0) = 4, \quad y'(0) = -5$$

Since it involves constant coefficients with initial values, we can use a Laplace Transform.

$$\mathcal{L}\{y''\} + 3\mathcal{L}\{y'\} + 2\mathcal{L}\{y\} = 0$$

$$(s^2Y(s) - sy(0) - y'(0)) + 3(sY(s) - y(0)) + 2Y(s) = 0$$

$$s^2Y - 4s + 5 + 3sY - 12 + 2Y = 0$$

We can solve the algebraic equation for $Y(s)$ so we can take the inverse Laplace Transform to get back to $y(t)$.

$$(s^2 + 3s + 2)Y = 4s + 7$$

$$Y(s) = \frac{4s+7}{(s+1)(s+2)}$$

$$\frac{4s+7}{(s+1)(s+2)} = \frac{A}{s+1} + \frac{B}{s+2}$$

$$4s + 7 = A(s+2) + B(s+1)$$

$$4s + 7 = (A+B)s + (2A+B)$$

$$A + B = 4, \quad 2A + B = 7 \Rightarrow A = 3, \quad B = 1$$

$$Y(s) = \frac{3}{s+1} + \frac{1}{s+2}$$

We can use the Laplace Transform rules in the table at the

beginning of the book to find that

$$y(t) = 3e^{-t} + e^{-2t}$$

Which is our final answer.

To summarize the steps,

1. Take Laplace Transform of both sides
2. Plug in initial conditions
3. Solve algebraically for $Y(s)$
4. Use partial fractions if needed
5. Take the inverse Laplace transform

5.3 Reduction of Order

If you have a third order or higher differential equation, we can use a method known as reduction of order to solve it. The biggest downfall is that it requires knowing at least one solution to the differential equation.

Reduction of Order

Given a known solution to a differential equation, y_1, then another solution is found by taking

$$y_2 = v(x)y_1 \tag{5.16}$$

Essentially, you plug this into the differential equation, and then your differential equation in terms of y will drop in order

5.4 Examples

Example 1

$$x^3y''' - 3x^2y'' + 6xy' - 6y = 0 \tag{5.17}$$

You can quickly verify that the function $y_1 = x$ is a valid solution to this differential equation, so we can use reduction of order. We make the substitution

$$y_2 = vx \tag{5.18}$$

$$y_2' = v'x + v \tag{5.19}$$

$$y_2'' = v''x + 2v' \tag{5.20}$$

$$y_2''' = v'''x + 3v'' \tag{5.21}$$

If we plug these back in, we get

$$x^3(v'''x + 3v'') - 3x^2(v''x + 2v') + 6x(v'x + v) - 6(vx) = 0 \tag{5.22}$$

Expanding this out, in order to cancel out the like terms, we get

$$x^4v''' + 3x^3v'' - 3x^3v'' - 6x^2v' + 6x^2v' + 6vx - 6vx = 0 \tag{5.23}$$

This drastically simplifies down to

$$x^4v''' = 0 \tag{5.24}$$

Now, we restrict $x \neq 0$ and divide by x^4 to both sides. This implies that

$$v''' = 0 \tag{5.25}$$

We are free to integrate this three times with respect to x, giving us

$$v = \frac{1}{2}C_1x^2 + C_2x + C_3 \tag{5.26}$$

We can multiply this by x to find the general solution to the differential equation.

$$y = \frac{1}{2}C_1x^3 + C_2x^2 + C_3x \tag{5.27}$$

Example 2

Solve

$$(y''' - y)^3 = e^{2x} \tag{5.28}$$

To solve this, we first take the cube root on both sides to force linearity. We can safely do this since there are no domain restrictions when the right-hand side is strictly positive.

$$y''' - y = e^{\frac{2}{3}x} \tag{5.29}$$

We can use undetermined coefficients to solve this. First, we solve the homogeneous equation.

$$y''' - y = 0 \tag{5.30}$$

$$r^3 - 1 = 0 \tag{5.31}$$

$$\Rightarrow r = 1, \frac{-1}{2} \pm \frac{\sqrt{3}}{2}i \tag{5.32}$$

Thus, the homogeneous solution is

$$y_h(x) = C_1 e^x + e^{-\frac{x}{2}}(C_2 \cos(\frac{\sqrt{3}}{2}x) + C_3 \sin(\frac{\sqrt{3}}{2}x)) \tag{5.33}$$

Now, to solve for the particular solution, we try the ansatz

$$y_p = Ae^{\frac{2}{3}x} \tag{5.34}$$

$$\Rightarrow y''' = A(\frac{2}{3})^3 e^{\frac{2}{3}x} = \frac{8}{27}Ae^{\frac{2}{3}x} \tag{5.35}$$

Substitute this back into the ODE.

$$\frac{8}{27}Ae^{\frac{2}{3}x} - Ae^{\frac{2}{3}x} = e^{\frac{2}{3}x} \tag{5.36}$$

Solving for A, we find that

$$A = -\frac{27}{19} \tag{5.37}$$

Therefore,

$$y_p = -\frac{27}{19}e^{\frac{2}{3}x} \tag{5.38}$$

If we combine this with the homogeneous equation, then we will find our final general solution to the ODE.

$$y = C_1e^x + e^{-\frac{x}{2}}(C_1\cos(\frac{\sqrt{3}}{2}x) + C_2\sin(\frac{\sqrt{3}}{2}x)) - \frac{27}{19}e^{\frac{2}{3}x} \tag{5.39}$$

Example 3

Solve

$$y''' - 2y + y = 0 \tag{5.40}$$

With the initial conditions

$$y''(0) = 0 \quad y'(0) = 0 \quad y(0) = 1 \tag{5.41}$$

This is set up extremely well for a Laplace transform due to the initial conditions, constant coefficients, and homogeneity. So, we apply the Laplace transform to both sides of the differential equation

$$\mathcal{L}\{y'''-2y''+y\} = s^3Y(s)-s^2-2(s^2Y(s)-s)+Y(s) \tag{5.42}$$

$$\mathcal{L}\{y'''-2y''+y\} = s^3Y(s)-s^2-2s^2Y(s)+2s+Y(s) \tag{5.43}$$

$$\mathcal{L}\{y''' - 2y'' + y\} = (s^3 - 2s^2 + 1)Y(s) + (-s^2 + 2s) \tag{5.44}$$

Next, we have to solve for $Y(s)$ in order to neatly take the inverse Laplace transform.

$$Y(s) = \frac{s^2 - 2s}{s^3 - 2s^2 + 1} \tag{5.45}$$

Factoring the denominator,

$$s^3 - 2s^2 + 1 = (s - 1)(s^2 - s - 1) \tag{5.46}$$

$$\Rightarrow Y(s) = \frac{s(s - 2)}{(s - 1)(s^2 - s - 1)} \tag{.5.47}$$

In order to split this up, we have to make use of partial

fraction decomposition. We let

$$\frac{s(s-2)}{(s-1)(s^2-s-1)} = \frac{A}{s-1} + \frac{Bs+C}{s^2-s-1} \tag{5.48}$$

Solving this for each of the unknown constants gives

$$A = -1, \quad B = 2, \quad C = 1 \tag{5.49}$$

Plugging these back in, we get

$$Y(s) = \frac{-1}{s-1} + \frac{2s+1}{s^2-s-1} \tag{5.50}$$

Since $Y(s)$ is in its most simplified form, we can take the inverse Laplace transform.
The first term is straightforward.

$$\mathcal{L}^{-1}\left\{\frac{-1}{s-1}\right\} = -e^t \tag{5.51}$$

For the second term, we make note the roots of the denominator.

$$s^2 - s - 1 = 0 \quad \Rightarrow \quad s = \frac{1 \pm \sqrt{5}}{2} \tag{5.52}$$

So the inverse transform is:

$$\mathcal{L}^{-1}\left\{\frac{2s+1}{s^2-s-1}\right\} = Ae^{\frac{1+\sqrt{5}}{2}t} + Be^{\frac{1-\sqrt{5}}{2}t} \tag{5.53}$$

So, we arrive at our final answer to the differential equation.

$$y(t) = -e^t + Ae^{\frac{1+\sqrt{5}}{2}t} + Be^{\frac{1-\sqrt{5}}{2}t} \tag{5.54}$$

Example 4

Given that $y(t) = \frac{1}{t}$ is a solution, find the general solution to

$$2t^2y'' + ty' - 3y = 0 \tag{5.55}$$

Since we were given one of the solutions, we can make use of reduction of order. So, we set

$$y_2 = vt^{-1} \tag{5.56}$$

Taking the first and second derivatives, we get

$$y_2' = -t^{-2}v + t^{-1}v' \tag{5.57}$$

$$y_2'' = 2t^{-3}v - 2t^{-2}v' + t^{-1}v \tag{5.58}$$

Substituting this back in, we get

$$2t^2(2t^{-3}v - 2t^{-2}v' + t^{-1}v) + t(-t^{-2}v + t^{-1}v') - 3vt^{-1} = 0 \tag{5.59}$$

If you multiply this out and cancel the like terms, you are left with

$$2tv'' - 3v' = 0 \tag{5.60}$$

Next, we make the substitution

$$w = v' \tag{5.61}$$

$$w = v'' \tag{5.62}$$

This transforms the differnetial equation into a first-order separable one.

$$2tw' - 3w = 0 \tag{5.63}$$

$$\frac{3}{2t} = \frac{w'}{w} \tag{5.64}$$

Integrating on both sides,

$$\int \frac{3}{2t}\, dt = \int \frac{w'(t)}{w(t)}\, dt \tag{5.65}$$

$$\frac{3}{2}\ln t = \ln w + C \tag{5.66}$$

$$\ln(t^{\frac{3}{2}}) = \ln w + C \Rightarrow w = C_1 t^{\frac{3}{2}} \tag{5.67}$$

Now, we substitute $w = v'$ back in to get

$$v' = C_1 t^{\frac{3}{2}} \tag{5.68}$$

Integrating this, we get

$$v = \frac{2}{5} C_1 t^{\frac{5}{2}} + C_2 \tag{5.69}$$

Lastly, if we plug this into the general form for the solution and multiply the terms, then we get the final answer.

$$y(t) = \frac{C_1}{t} + C_2 t^{\frac{3}{2}} \tag{5.70}$$

Example 5

Solve

$$y''' = e^y \tag{5.71}$$

Given the differential equation

$$y''' = -e^y, \tag{5.72}$$

we try the ansatz

$$y = \ln(Cx^r) = \ln(C) + r\ln(x). \tag{5.73}$$

First, we compute the derivatives of y. The first derivative is

$$y' = \frac{r}{x}. \tag{5.74}$$

The second derivative is

$$y'' = -\frac{r}{x^2}. \tag{5.75}$$

The third derivative is

$$y''' = \frac{2r}{x^3}. \tag{5.76}$$

Next, we evaluate the right-hand side of the equation. Since

$$e^y = e^{\ln(Cx^r)} = Cx^r, \tag{5.77}$$

we have

$$-e^y = -Cx^r. \tag{5.78}$$

Substituting these expressions into the differential equation gives

$$\frac{2r}{x^3} = -Cx^r. \tag{5.79}$$

For this equality to hold for all x, the powers of x must be equal. Therefore,

$$-3 = r, \tag{5.80}$$

which implies

$$r = -3. \tag{5.81}$$

Substituting $r = -3$ back into the equation, we obtain

$$\frac{2 \cdot (-3)}{x^3} = -Cx^{-3}. \tag{5.82}$$

This simplifies to

$$\frac{-6}{x^3} = -Cx^{-3}. \tag{5.83}$$

Equating coefficients, we find

$$-6 = -C, \tag{5.84}$$

which leads to

$$C = 6. \tag{5.85}$$

Hence, the solution to the differential equation is

$$y = \ln(6x^{-3}) = \ln(6) - 3\ln(x). \tag{5.86}$$

Chapter 6

Special Equations

SPECIAL DIFFERENTIAL EQUATIONS show up frequently enough and in different fields that they have been named by the people who first derived them or first solved them. Beyond the following equations, there are many more examples of differential equations that have proved useful for math, physics, and more. This chapter serves to give you a selection of this.

6.1 Bernoulli Equation

The first differential equation we will look at is the Bernoulli Equation.

Bernoulli Equation

The Bernoulli Equation is given by

$$\frac{dy}{dx} + P(x)y = Q(x)y^n \tag{6.1}$$

Where $n \in \Re$, and both $P(x)$ and $Q(x)$ are real-valued functions.

In order to solve this, we make the substitution

$$v = y^{1-n} \tag{6.2}$$

We take the derivative of this with respect to x.

$$\frac{dv}{dx} = (1-n)y^{-n}\frac{dy}{dx} \Rightarrow \frac{dy}{dx} = \frac{1}{1-n}y^n\frac{dv}{dx} \tag{6.3}$$

Now, we can substitute this into the original differential equation.

$$\frac{1}{1-n}y^n\frac{dv}{dx} + P(x)y = Q(x)y^n \tag{6.4}$$

We divide both sides of this by y^n.

$$\frac{1}{1-n}\frac{dv}{dx} + P(x)y^{1-n} = Q(x) \tag{6.5}$$

Since $v = y^{1-n}$, we can substitute it in.

$$\frac{1}{1-n}\frac{dv}{dx} + P(x)v = Q(x) \tag{6.6}$$

For the last step, we multiply both sides by $(1 - n)$.

$$\frac{dv}{dx} + (1-n)P(x)v = Q(x)(1-n) \tag{6.7}$$

Recognize that this is a linear first-order differential equation that can be solved using an integrating factor or using the general formula.

6.2 Clairaut's Equation

Clairaut's equation is a first-order non-linear differential equation. What makes it unique is the $f(\frac{dy}{dx})$ term, something that is rarely seen.

Clairaut's Equation

Clairaut's equation is given by

$$y = x\frac{dy}{dx} + f(\frac{dy}{dx}) \tag{6.8}$$

In order to solve it, we transform $\frac{dy}{dx} \to p$. This turns the differential equation into an algebraic one.

$$y = xp + f(p) \tag{6.9}$$

This is a straight line. Who cares? Well, the family of straight lines for different values of p is known as the envelope. The interesting property is that each line is tangent to the actual solution in at least one point. Once we have the envelope, we can solve for a singular solution. Consider the following case

$$y = x\frac{dy}{dx} + \ln(\frac{dy}{dx}) \tag{6.10}$$

The envelope is given by

$$y = xp + \ln p \tag{6.11}$$

To find the singular solution, we differentiate the envelope with respect to p.

$$0 = x + \frac{1}{p} \Rightarrow x = \frac{-1}{p} \Rightarrow p = \frac{-1}{x} \tag{6.12}$$

We can now substitute $x = \frac{-1}{p}$ into the envelope.

$$y = (\frac{-1}{p})p + \ln p \tag{6.13}$$

$$y = -1 + \ln p \tag{6.14}$$

Lastly, we substitute in $p = \frac{-1}{x}$.

$$y = -1 + \ln(\frac{-1}{x}) \tag{6.15}$$

$$y = -1 - \ln(-x) \quad \text{for } x < 0 \tag{6.16}$$

This can be extended for when we have other functions as well. If instead we had $(\frac{dy}{dx})^2$ or $\sin(\frac{dy}{dx})$, we would still handle the problem in the same way.

Consider a few examples of what this looks like.

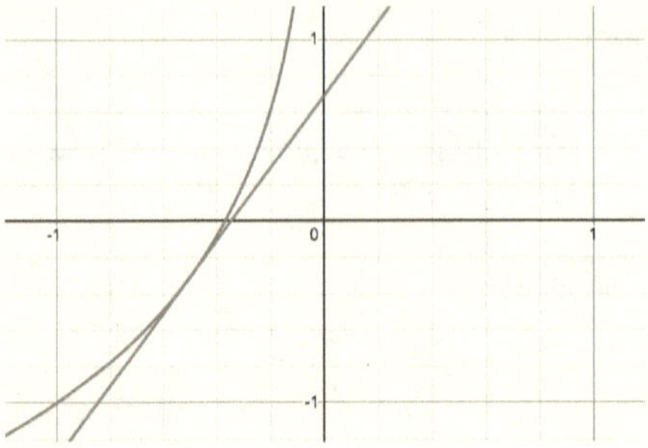

Figure 6.1: $y(x)$ and envelope at $p = 3$

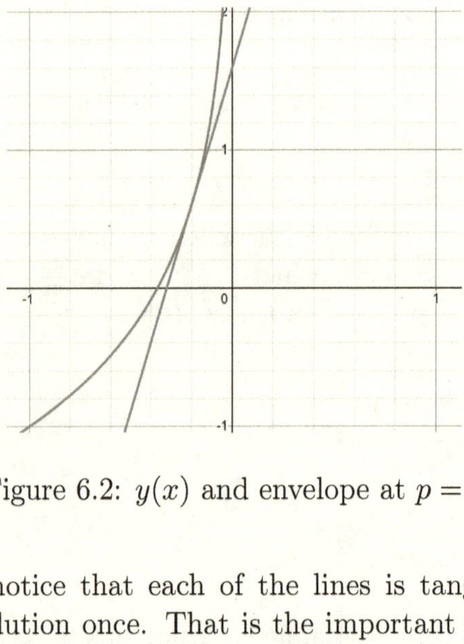

Figure 6.2: $y(x)$ and envelope at $p = 5$

You may notice that each of the lines is tangent to the singular solution once. That is the important property of this equation that makes it so notable.

6.3 Euler-Cauchy Equation

The Euler-Cauchy equation is a second-order homogeneous equation with variable coefficients. It is given with the general coefficients:

Euler-Cauchy Equation

$$x^2y'' + axy' + by = 0 \qquad (6.17)$$

6.4 Bessel's Equation

Bessel's equation is a second-order linear ODE that often pops up in cylindrical or spherical symmetry. Oftentimes, it will arise due to vibrations, heat conduction, and even wave propagation.

Bessel Equation

$$x^2y'' + xy' + (x^2 - \nu^2)y = 0 \tag{6.18}$$

The first solution, which is found by using a power series, is given by

$$J_\nu(x) = \sum_{m=0}^{\infty} \frac{(-1)^m}{m!\Gamma(m+\nu+1)} (\frac{x}{2})^{2m+\nu} \tag{6.19}$$

This converges for all real values of x. ν is the order of the equation. For integer values of ν,

$$J_{-\nu}(x) = (-1)^\nu J_\nu \tag{6.20}$$

The second solution to the differential equation is given in terms of J_ν.

$$Y_\nu(x) = \frac{J_\nu(x)\cos(\nu\pi) - J_{-\nu}(x)}{\sin(\nu\pi)} \quad \text{for} \quad \nu \notin \mathbb{Z} \tag{6.21}$$

6.5 Airy's Equation

Airy's differential equation is one that looks deceivingly simple. It appears in specific instances in quantum mechanics, along with optics.

Airy Equation

$$y'' - xy = 0 \tag{6.22}$$

This equation can be solved using a power series, however it is more common to write the two linearly independent solutions in the form of two integrals.

$$y = C_1\mathrm{Ai}(x) + C_2\mathrm{Bi}(x) \tag{6.23}$$

Where

$$\mathrm{Ai}(x) = \frac{1}{\pi}\int_0^\infty \cos(3t^3 + xt)dt \tag{6.24}$$

And

$$\mathrm{Bi}(x) = \frac{1}{\pi}\int_0^\infty \exp(\frac{-t^3}{3} + xt) + \sin(3t^3 + xt)dt \tag{6.25}$$

6.6 Sturm-Louiville Form

The Sturm-Louiville form is a specific class of second-order linear differential equation. They are widely used in mathematical physics, especially in problems that make use of eigenvalues and orthogonality.

Sturm-Louisville Form

The Sturm-Louiville form of a differential equation is given by

$$\frac{d}{dx}(p(x)\frac{dy}{dx}) + (\lambda w(x) - q(x))y = 0 \qquad (6.26)$$

In this, $p(x)$, $q(x)$, and $w(x)$ are all real-valued functions. $w(x)$ has its own special name. It is known as the weight function. λ is an eigenvalue corresponding to each problem. In order to turn this into a Sturm-Louiville problem, we impose the following boundary conditions.

$$a_1 y(c) + a_2 y'(c) = 0 \qquad b_1 y(d) + b_2 y'(d) = 0 \qquad (6.27)$$

If we have a general differential equation, we can convert it into Sturm-Louisville form.

$$a_2(x)y'' + a_1(x)y' + a_0 y = \lambda y \qquad (6.28)$$

We make use of an integrating factor, $\mu(x)$, so that

$$\mu(x)a_2(x)y'' + \mu(x)a_1(x)y' = \frac{d}{dx}(\mu(x)a_2(x)y') \qquad (6.29)$$

This integrating factor is found by taking

$$\mu(x) = \frac{1}{a_2(x)} \exp(\int \frac{a_1(x)}{a_2(x)}\, dx) \tag{6.30}$$

We find the weight function by dividing this by equation by $a_2(x)$.

$$w(x) = \frac{\mu(x)}{a_2(x)} = \frac{1}{(a_2(x))^2} \exp(\int \frac{a_1(x)}{a_2(x)}\, dx) \tag{6.31}$$

This form can be applied to many different types of differential equations, three of which we will look at in the following sections. They have all three proved important in heat-flow analysis, quantum mechanics, and wave analysis.

We can write solutions to Sturm-Louisville differential equations using polynomials. The solutions to the Legrendre, Laguerre, and Hermite equations will all be in terms of polynomials since they can be rewritten to match the Sturm-Louisville form.

Rodrigues Formula

The Rodrigues formula is given by

$$P_n = \frac{1}{K_n w(x)} \frac{d^n}{dx^n} (w(x) p^n(x)) \tag{6.32}$$

As you plug in values of n, this will spit out polynomials that solve the Sturm-Lousiville differential equation for whatever functions $w(x)$ and $p(x)$ you have.

6.7 Legendre's Equation

Legendre's differential equation is useful when solving spherical PDEs. The Laplace equation:

$$\nabla^2 u = 0 \tag{6.33}$$

which appears in electrostatics, gravitational potentials, fluid-flow, and steady-state heat conduction will simplify down to Legendre's equation in spherical coordinates.

Legendre's Equation

$$(1 - x^2)y'' - 2xy' + n(n + 1)y = 0 \tag{6.34}$$

You may immediately notice that, using the product rule,

$$\frac{d}{dx}((1 - x^2)y') = (1 - x^2)y'' - 2xy' \tag{6.35}$$

So, we can rewrite the differential equation as

$$\frac{d}{dx}((1 - x^2)y') + n(n + 1) = 0 \tag{6.36}$$

If we compare this to our Rodrigues formula, then we would see that the weight function, $w(x) = 1$ and $p(x) = 1 - x^2$. Therefore, the formula for the polynomials, P_n, can be written as

$$P_n(x) = \frac{1}{K_n}\frac{d^n}{dx^n}((1 - x^2)^n) \tag{6.37}$$

If we consider what the order of the coefficients is doing as $n \to \infty$, we see they are on the order of $2^n n!$. This is an issue because the size of the coefficients will grow too fast.

So, to counteract this, we let $K_n = 2^n n!$. Therefore,

$$P_n = \frac{1}{2^n n!} \frac{d^n}{dx^n}((1 - x^2)^n) \tag{6.38}$$

Often times, textbooks will actually factor out a $(-1)^n$ term in order to flip the order of $(1 - x^2)$.

$$P_n = \frac{(-1)^n}{2^n n!} \frac{d^n}{dx^n}((x^2 - 1)^n) \tag{6.39}$$

6.8 Laguerre's Equation

Laguerre's equation is useful for solving the radial part of the hydrogen atom using Schrodinger's equation. The solution, a wave function, says how likely you are to find the electron at a specific radius (in this case), given the energy level of the atom. It can also be used to numerically approximate certain integrals of the form $\int_0^\infty f(x)e^{-x}\,dx$.

Laguerre's Equation

$$xy'' + (1-x)y' + n(n+1)y = 0 \tag{6.40}$$

In this case, there isn't a quick simplification we can make, so we will define the integrating factor.

$$\mu(x) = \frac{1}{x}\exp(\int \frac{1-x}{x}\,dx) \tag{6.41}$$

$$\mu(x) = \frac{1}{x}e^{\ln x}e^{-x} \tag{6.42}$$

$$\mu(x) = e^{-x} \tag{6.43}$$

We multiply this through the differential equation.

$$xe^{-x}y'' + (1-x)e^{-x}y' + ne^{-x}y = 0 \tag{6.44}$$

Now we recognize that

$$\frac{d}{dx}(xe^{-x}y') + ne^{-x}y = 0 \tag{6.45}$$

is equivalent via the product rule. So, the Rodrigues formula is

$$\frac{e^x}{n!}\frac{d}{dx}(e^{-x}x^n) \tag{6.46}$$

6.9 Hermite's Equation

The last special differential equation that we will look at is the Hermite equation. It is primarily used in quantum mechanics when solving the time-independent Schrodinger equation for the quantum harmonic oscillator. It can also be used for approximating probability distributions, specifically Gaussian distributions.

Hermite's Equation

$$y'' - 2xy' + 2ny = 0 \tag{6.47}$$

Once again, we define our integrating factor.

$$\mu(x) = \exp\left(\int (-2x)\, dx\right) = e^{-x^2} \tag{6.48}$$

Therefore,

$$e^{-x^2}y'' - 2xe^{-x^2}y' + 2ne^{-x^2}y = 0 \tag{6.49}$$

$$\frac{d}{dx}(e^{-x^2}y') + 2ne^{-x^2}y = 0 \tag{6.50}$$

Therefore, our Rodriguues formula is

$$H_n(x) = (-1)^n e^{x^2}\frac{d^n}{dx^n}(e^{-x^2}) \tag{6.51}$$

Due to the e^{-x^2}, you can see why this relates to the Gaussian distribution. For this reason, it also can help approximate

$$\int_0^\infty f(x)e^{-x^2}\, dx \tag{6.52}$$

Chapter 7

Numerical Methods

Numerical methods are beneficial when we are given a differential equation that applies to a real-world situation; however, the problem is not always solvable using analytic techniques. Sometimes, it may even be more important to have a qualitative description of a system (oscillatory, decreasing, increasing, stable) rather than the exact solution. Fluids, air, electrical circuits, and structural analysis are all examples of situations where the differential equations produced may only be solved numerically. For this reason, mathematicians have come up with a plethora of methods to approximate the solution to a differential equation based on characteristics.

Numerical methods work by discretizing the differential equation into uniform steps n over time and/or space. A few of the methods we see will refer to previous states of the differential equation that influence the current step. Then, you can build an iterative pattern that allows you to predict the value of the solution to the differential equation far from the original starting position.

7.1 Forward Euler Method

The Forward Euler method, or simply the Euler method, is the most fundamental numerical technique for solving first-order differential equations. Using it, we can approximate

$$\frac{dy}{dt} = f(t, y), \qquad y(t_0) = t_0 \tag{7.1}$$

It uses the iterative formula

$$y_{n+1} = y_n + hf(t_n, y_n) \tag{7.2}$$

Where y_n is the current approximation of the solution y at time t_n, h is the step size, and $f(t_n, y_n)$ is the value of the derivative at the current point. The next time point is found by taking

$$t_{n+1} = t_n + h \tag{7.3}$$

Throughout this chapter, we will make use of the following differential equation

$$\frac{dy}{dt} = t + y, \qquad y(0) = 1 \qquad \text{with } h = 0.1 \tag{7.4}$$

The table below gives each step with the values in between, up to $y(1.3)$.

Compared to the methods we will use later, this is generally easy to implement; however, it is not suitable for stiff equations. A stiff equation means that the solution is numerically unstable unless one takes an extremely small step.

Step n	t_n	y_n	$f(t_n, y_n) = t_n + y_n$	y_{n+1}
0	0.0	1.000	1.000	1.100
1	0.1	1.100	1.200	1.220
2	0.2	1.220	1.420	1.362
3	0.3	1.362	1.662	1.528

If you were to try to approximate $y(10)$ using this method, the error would be astronomical due to compounding differences.

7.2 Backward Euler Method

The Backward Euler method, also known as the Implicit Euler method, is a technique that is very similar to its forward counterpart. The major difference is that the approximation relies on the current point rather than on previous points. Given

$$\frac{dy}{dt} = f(t, y), \qquad y(t_0) = t_0 \tag{7.5}$$

The next point is calculated using

$$y_{n+1} = y_n + hf(t_{n+1}, y_{n+1}) \tag{7.6}$$

For simple differential equations, it is relatively easy to solve them; however, if random terms of y and functions of it are involved, it becomes complex and challenging. Other methods can perform much better, especially if you are using a program to compute y. It does have its limited uses; however, it falls into the same trap as the forward version; it cannot deal with stiff differential equations. For our example,

$$y_{n+1} = y_n + h(t_{n+1} + y_{n+1}) \tag{7.7}$$

Now, we can solve for y_{n+1}.

$$y_{n+1} - hy_{n+1} = y_n + ht_{n+1} \tag{7.8}$$

$$y_{n+1} = \frac{y_n + ht_{n+1}}{1 - h} \tag{7.9}$$

Again, below is a table of values calculated up to $t = 0.3$.

Step n	t_n	y_n	t_{n+1}	y_{n+1}
0	0.0	1.0000	0.1	1.1222
1	0.1	1.1222	0.2	1.2691
2	0.2	1.2691	0.3	1.4435
3	0.3	1.4435	0.4	1.6483

7.3 Midpoint Method

The midpoint method is a two step process. It involves calculating a value at the midpoint of y_n and y_{n+1}. Then, you use the value at $y_{n+\frac{1}{2}}$ to calculate the next point. The formula is as follows:

$$k_1 = f(t_n, y_n) \tag{7.10}$$

$$y_{n+\frac{1}{2}} = y_n + \frac{h}{2}k_1 \tag{7.11}$$

Then, step two estimates y_{n+1}.

$$k_2 = f(t_n + \frac{h}{2}, y_{n+\frac{1}{2}}) \tag{7.12}$$

$$y_{n+1} = y_n + hk_2 \tag{7.13}$$

Going back to the example, the table below provides the approximations.

n	t_n	y_n	k_1	$y_{n+\frac{1}{2}}$	k_2	y_{n+1}
0	0.0	1.00000	1.00000	1.05000	1.10000	1.11000
1	0.1	1.11000	1.21000	1.17050	1.32050	1.24205
2	0.2	1.24205	1.44205	1.31415	1.56415	1.39847

7.4 Runge-Kutta Method (RK4)

The Runge-Kutta (RK4) method is an extension of the midpoint method. It uses four steps to calculate $n + 1$, which are given by

$$\begin{aligned}
k_1 &= f(t_n, y_n) \\
k_2 &= f\left(t_n + \frac{h}{2},\ y_n + \frac{h}{2}k_1\right) \\
k_3 &= f\left(t_n + \frac{h}{2},\ y_n + \frac{h}{2}k_2\right) \\
k_4 &= f\left(t_n + h,\ y_n + hk_3\right) \\
y_{n+1} &= y_n + \frac{h}{6}\left(k_1 + 2k_2 + 2k_3 + k_4\right)
\end{aligned}$$

RK4 is the most popular choice as it is extremely accurate. The more steps you use, the more intensive each iteration is; however, four is a happy medium between speed and accuracy. Below is a table for the recurring example.

n	t_n	y_n	k_1	k_2	k_3	k_4	y_{n+1}
0	0.0	1.0000	1.00000	1.10000	1.1050	1.2105	1.1103
1	0.1	1.1103	1.2103	1.3208	1.3263	1.4439	1.2428
2	0.2	1.2428	1.4428	1.5649	1.5710	1.6999	1.4004

7.5 Backwards Differentiation

Backwards Differentiation Formulas (BDFs) are implicit multistep methods that solve stiff ODEs. The simplest BDF, BDF1, is simply the Backward Euler method. BFD2 is given by

$$y_{n+1} = \frac{4}{3}y_n - \frac{1}{3}y_{n-1} + \frac{2}{3}hf(t_{n+1}, y_{n+1}) \tag{7.14}$$

Below is a table of BDF1 through BDF5 based on their coefficients.

	Backward Differentiation Formula (BDF)
1	$y_{n+1} - y_n = hf(t_{n+1}, y_{n+1})$
2	$\frac{3}{2}y_{n+1} - 2y_n + \frac{1}{2}y_{n-1} = hf(t_{n+1}, y_{n+1})$
3	$\frac{11}{6}y_{n+1} - 3y_n + \frac{3}{2}y_{n-1} - \frac{1}{3}y_{n-2} = hf(t_{n+1}, y_{n+1})$
4	$\frac{25}{12}y_{n+1} - 4y_n + 3y_{n-1} - \frac{4}{3}y_{n-2} + \frac{1}{4}y_{n-3} = hf(t_{n+1}, y_{n+1})$

Comparing the results for each of these techniques, we see that Runge-Kutta (RK4) is the most accurate by far.

t	Exact	Forward	Backward	Midpt	RK4	BDF2
0.1	1.11034	1.1000	1.12222	1.11000	1.11034	0.00000
2.0	11.77811	10.4550	13.45053	11.73247	11.77809	1.31309

The error between the exact solution and RK4 after 20 iterations is only 0.00002, which represents a percent error of $1.698 \times 10^{-4}\%$.

Chapter 8

Applications

DIFFERENTIAL EQUATIONS ARE HIGHLY APPLICABLE. As we discussed previously in chapter 1, differential equations are crucial to many different fields. In this chapter, we seek to expand upon these applications and actually solve the differential equations while applying them to specific problems. This chapter will not cover every use of differential equations. There are almost limitless problems that can be solved using differential equations.

8.1 Harmonic Motion

Newton examined different physical systems and used differential equations to model them. From planetary motion to gravitation to springs, his models still stand true today. The first differential equation we will examine describes how a mass attached to a spring moves under the influence of restoring forces or damping. A restorative force is any force that acts in opposition to motion. In other words, a restorative force is a force that aims to establish static equilibrium.

Consider a mass m attached to a spring with a spring constant k.

Hooke's Law

Hooke's Law states that the restoring force exerted by a spring on a mass is

$$F = -kx \tag{8.1}$$

where x is the displacement from equilibrium and k is the constant associated with the spring.

The spring constant, k, depends on the size of the spring, the material of which it is made, and other factors. You may also recall Newton's Second Law

$$F = ma \tag{8.2}$$

or, written as a differential equation in terms of position,

$$F = m\frac{d^2x}{dt^2} \tag{8.3}$$

We can equate the two, giving

$$m\frac{d^2x}{dt^2} + kx = 0 \tag{8.4}$$

This is a second-order linear homogeneous differential equation. We can solve it by writing the characteristic equation.

$$mr^2 + k = 0 \tag{8.5}$$

$r = \pm i\sqrt{\frac{k}{m}}$ So, we write our solution to the differential equation.

$$x(t) = A\cos(\sqrt{\frac{k}{m}}t) + B\sin(\sqrt{\frac{k}{m}}t) \tag{8.6}$$

The constants A and B are determined by the initial conditions. The intuitive idea is that when the spring is furthest from equilibrium, it is at its maximum amplitude. This is the value of A.

Now, if we add damping to the system via air resistance or friction, the differential equation also depends on velocity.

$$m\frac{d^2x}{dt^2} + c\frac{dx}{dt} + kx = 0 \tag{8.7}$$

This is still a linear-homogeneous equation. If you were to solve it, you would find three cases depending on the value of the discriminant, $c^2 - 4mk$.

If $c^2 - 4mk < 0$, then the spring is underdamped. In this case, the spring is allowed to oscillate back and forth. It returns to equilibrium slowly.

If $c^2 - 4mk > 0$, then the spring is overdamped. It will not oscillate and will slowly return to equilibrium.

Lastly, if $c^2 - 4mk = 0$, the spring is critically damped. This will result in the shortest amount of time it takes for the spring to reach equilibrium with no oscillations.

Combining this situation with forced oscillations, we can apply a force to our mass while it is in motion and observe how that affects the displacement. Instead of going to 0, the right-hand side is given by the force term $F_0 \cos(\gamma t)$, where γ is the frequency of the applied force. This is the basis for understanding resonance. Resonance is when a system is driven at its natural frequency and exhibits large oscillations. It is due to constructive interference amplifying the original oscillations. This differential equation is non-homogeneous and is given by

$$m\frac{d^2x}{dt^2} + c\frac{dx}{dt} + kx = F_0 \cos(\gamma t) \tag{8.8}$$

This has great applications in structural analysis. The Millennium Bridge in London became famous for its wobbliness. This was caused by the natural frequency aligning with the vibrations caused by the pedestrians' steps. For this reason, the amplitude of the swaying back and forth was amplified.

Natural frequencies are also the reason why, if you sing

at the right frequency, you may be able to break certain glasses. The sound waves travel at a certain frequency that interacts with the glass's frequency, causing small vibrations to be magnified and resulting in a break.

8.2 Population Biology

Differential equations have been used to model how populations grow and change over time. The simplest model of population growth is exponential. This assumes that resources are unlimited and that the organisms can reproduce endlessly and never die. Obviously, this is unrealistic. The differential equation is given by

$$\frac{dN}{dt} = rN \tag{8.9}$$

N is the population at time t, and r is the rate at which the organisms are reproducing. This is the first order and can be separated to find that

$$N(t) = Ce^{rt} \tag{8.10}$$

C is the initial population N_0, which we establish by setting $t = 0$.

More realistically, we have the logistic-growth model. This incorporates additional environmental limitations such as resource limits, predation, space, and competition. It is described by the following first-order nonlinear differential equation.

$$\frac{dN}{dt} = rN(1 - \frac{N}{K}) \tag{8.11}$$

K is what is known as the carrying capacity. This tells us how many organisms can be alive and reproducing before they run out of resources. We can use separation of variables to solve this.

$$\frac{dN}{dt} = rN(\frac{K-N}{K}) \tag{8.12}$$

$$\frac{K}{K-N}\ dN = r\ dt \tag{8.13}$$

We can integrate this on both sides; however, we first need to split this up. Add and subtract N from the numerator.

$$\int \frac{K-N+N}{N(K-N)}\ dN = rt + C \tag{8.14}$$

We can split the fraction into 2.

$$\int (\frac{K-N}{N(K-N)} + \frac{N}{N(K-N)})\ dN = rt + C \tag{8.15}$$

We can cancel the terms.

$$\int (\frac{1}{N} + \frac{1}{K-N})\ dN = rt + C \tag{8.16}$$

Now we can integrate each term.

$$\ln N - \ln(K-N) \ = rt + C \tag{8.17}$$

We need to now solve for $N(t)$. Using logarithm rules,

$$\ln(\frac{N}{K-N}) = rt + C \tag{8.18}$$

$$\frac{N}{K-N} = Ce^{rt} \tag{8.19}$$

$$N = Ce^{rt}(K-N) \tag{8.20}$$

$$N + Ne^{rt} = KCe^{r}t \tag{8.21}$$

$$N(1 + Ce^{rt}) = KCe^{rt} \tag{8.22}$$

$$N(t) = \frac{KCe^{rt}}{1 + Ce^{rt}} \tag{8.23}$$

Lastly, we divide by Ce^{rt} on the top and bottom to get

$$N(t) = \frac{K}{1 + Ce^{-rt}} \tag{8.24}$$

Graphing population vs. time, we obtain a sigmoid curve (s-shaped). It has rapid growth at the start, slow growth as it reaches its carrying capacity, and it stops growing entirely once it reaches $N = K$.

The logistic growth can be interpreted as encapsulating density-dependent interactions and growth, population equilibrium, and population stability. It can be used to model bacterial growth in a petri dish, animal populations in dynamical systems, human population growth, the spread of invasive species, and the management of overhunting, overfishing, and overharvesting. It can also be extended to account for seasonal changes, predator-prey interactions, and much more.

8.3 Reaction Rate

Differential equations can be used to model chemical reactions and how the concentrations of reactants vs. products change over time. For a simple, irreversible reaction

$$A \to B \tag{8.25}$$

The rate law determines how the concentration of the reactant $[A]$ changes over time. For a first-order reaction, the rate is proportional to the concentration of A.

$$\frac{dA}{dt} = -kA \tag{8.26}$$

k is the rate constant specific to the reaction and temperature.

We can separate this as follows.

$$\frac{1}{A}\,dA = -k\,dt \tag{8.27}$$

$$\ln A = -kt \tag{8.28}$$

$$A = A_0 e^{-kt} \tag{8.29}$$

For higher order reactions,

$$\frac{dA}{dt} = -kA^n \tag{8.30}$$

$$\frac{1}{A^n}\frac{dA}{dt} = -k \tag{8.31}$$

$$A^{-n}dA = -kt + C \tag{8.32}$$

$$-A^{1-n} = -kt + C \tag{8.33}$$

$$A^{1-n} = kt + C \tag{8.34}$$

$$A = (kt + C)^{1-n} \tag{8.35}$$

We can also combine this with a system of reactions and model their progression using differential equations.

$$A \to B \to C \tag{8.36}$$

$$\frac{dA}{dt} = -k_1 A \tag{8.37}$$

$$\frac{dB}{dt} = k_1 A - k_2 B \tag{8.38}$$

$$\frac{dC}{dt} = k_2 B \tag{8.39}$$

So long as the k coefficients are constant, this system can be solved analytically.

Reaction rates are useful in many fields. Pharmacokineticists use differential equations to determine the half-life of certain drugs in the human body in order to establish an appropriate dosage and interval. Many drugs metabolized in the liver follow first-order kinetics. Environmental chemists use reaction rates to determine how quickly pollutants will degrade in sunlight and when exposed to the elements. This allows us to determine the biodegradability of certain materials and compounds. Lastly, for industrial uses, chemists determine which combination of catalyst concentrations and temperature will yield the desired product output based on their targeted reaction rate. These differential equations have also been theoretically extended to account for reversible reactions, catalysis, and temperature dependence.

8.4 Economics

Differential equations can be used to model growth, interest, inflation, and returns on investment over time. Consider the Solow Growth Model. It describes how capital accumulates in an economy over time.

$$\frac{dK}{dt} = sY - \delta K \tag{8.40}$$

Y is the Cobb-Douglas production function

$$Y = K^{\alpha} L^{1-\alpha} \tag{8.41}$$

where L is the labor. s is the savings rate, and δ is the depreciation rate. Compounding interest rates are governed by the exact same differential equation as exponential population growth. Economists can use differential equations in many fascinating ways to make economic and market predictions. They can model risk and portfolio management.

8.5 Structural Analysis

If you look at cutting edge skyscrapers, you may wonder how exactly they remain standing. There is a lot of mathematics that go into constructing a structurally sound and safe building. One of the equations that structural engineers may use is the Euler-Bernoulli beam equation. It is a fourth order partial differential equation that describes how a slender beam will bend. The transverse deflection (displacement away from the stable axis), $w(x,t)$, is governed by the differential equation:

$$EI\frac{\partial^4 w(x,t)}{\partial x^4} + \rho A\frac{\partial^2 w(x,t)}{\partial t^2} = q(x,t) \tag{8.42}$$

E is the Young Modulus of the material. It essentially measures how stiff an object is. Higher values correspond to a greater resistance to bending. Same story for I. It is the second moment of a quantity that describes how the area is distributed concerning a specific arbitrary axis. ρ is the density of the material from which the beam is made, and A is its cross-sectional area. Lastly, $q(x,t)$ is the load distributed into the beam, causing it to bend.

This equation, due to its fourth-order, is generally difficult, if not impossible, to solve analytically except for specific cases. For example, if we assume the beam is static (does not depend on time), then we can eliminate the second time derivative and the time component of the displacement, converting this to an ODE.

$$EI\frac{d^4w}{dx^4} = q(x) \tag{8.43}$$

If we assume that $q(x) = q_0$ is a uniformly distributed load, with boundary conditions

$$w(0) = 0, \quad w(L) = 0, \quad w''(0) = 0, \quad w''(L) = 0 \tag{8.44}$$

So, we have

$$\frac{d^4w}{dx^4} = \frac{q_0}{EI} \tag{8.45}$$

We now integrate this four times.

$$w''' = \frac{q_0}{EI}x + C_1 \tag{8.46}$$

$$w'' = \frac{q_0}{2EI}x^2 + C_1x + C_2 \tag{8.47}$$

$$w' = \frac{q_0}{6EI}x^3 + \frac{1}{2}C_1x^2 + C_2x + C_3 \tag{8.48}$$

$$w = \frac{q_0}{24EI}x^4 + \frac{C_1}{6}x^3 + \frac{C_2}{2}x^2 + C_3x + C_4 \tag{8.49}$$

Now, we apply the initial conditions.

$$w(0) = 0 \Rightarrow C_4 = 0 \tag{8.50}$$

$$w''(0) = 0 \Rightarrow C_2 = 0 \tag{8.51}$$

$$w(L) = 0 \Rightarrow \frac{q_0}{24EI}L^4 + \frac{C_1}{6}L^3 = 0 \tag{8.52}$$

$$\Rightarrow C_1 = -\frac{q_0L}{4EI} \tag{8.53}$$

$$w''(L) = 0 \Rightarrow C_3 = \frac{q_0 L^3}{24} \tag{8.54}$$

Combining all of this together, we obtain the deflection of our beam.

$$w(x) = \frac{q_0}{24EI}(x^4 - 2Lx^3 + L^3x) \tag{8.55}$$

There are different cases that you can have for a beam's deflection. You can have a beam that is embedded in the wall at both ends $x = 0$ and $x = L$; you can have a beam that is simply supported at both ends. Lastly, you can have a cantilever beam, which means that it is embedded in one wall at $x = 0$ and left free to dangle. Each of these varies in their initial conditions, so the coefficients of the polynomial vary slightly. Consider the following graphs of how a beam of length 10 would deflect.

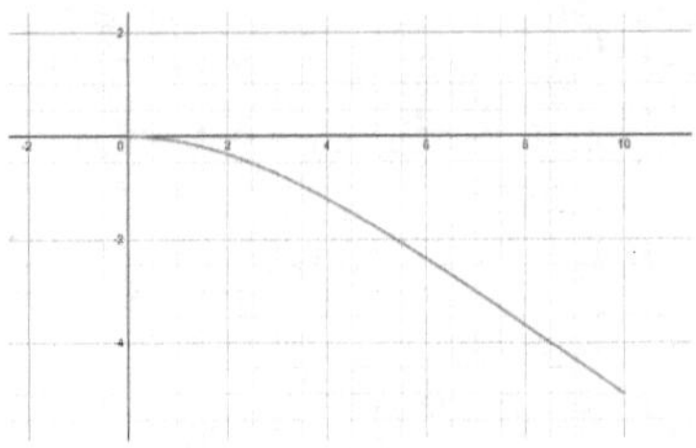

Figure 8.1: Cantilever Beam $y = -\frac{40}{240000}(x^4 - 40x^3 + 600x^2)$

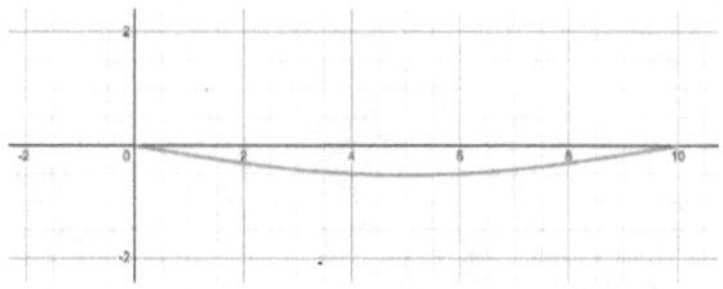

Figure 8.2: Simply-supported Beam $y = -\frac{40}{240000}(x^4 - 20x^3 + 1000x)$

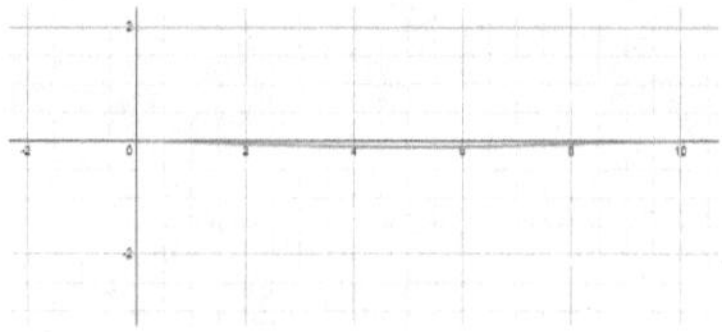

Figure 8.3: Embedded Beam $y = -\frac{40}{240000}(x^4 - 20x^3 + 100x^2)$

If you plug the following equations into any graphing calculator—I prefer Desmos—you can also show what happens if you apply a steady force to the beam at a specific point along its length, a.

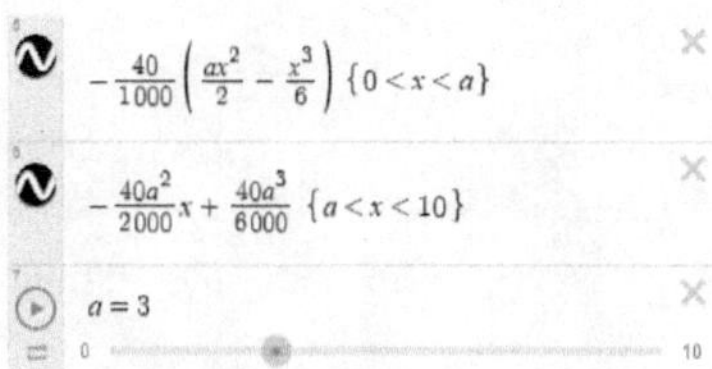

Figure 8.4: Equation for point-load

Chapter 9

Introduction to Partial Differential Equations

PARTIAL DIFFERENTIAL EQUATIONS are differential equations written in more than 1 variables. You may have spatial and time derivatives, something that we did not see when dealing with ordinary differential equations.

9.1 Terminology

We seek to answer three questions in this section.

1. What are they?

2. Why do we care?

3. What types are there?

We have already answered the first one, but what about the other two. Who cares about partial differential equations? Well, anyone who relies on modern technology should. Modern aerodynamics, architecture/engineering, heating/cooling, and even traffic can all be modeled using partial differential equations. Some of these we can solve analytically, but many we have to approximate numerically.

To answer the third question, there are three major types.

Classification of PDEs

Given the partial differential equation

$$a\frac{\partial^2 u}{\partial x^2} + b\frac{\partial^2 u}{\partial x \partial y} + c\frac{\partial^2 u}{\partial y^2} = 0 \tag{9.1}$$

A hyperbolic PDE has $b^2 - 4ac > 0$.
An elliptic PDE has $b^2 - 4ac < 0$.
A parabolic PDE has $b^2 - 4ac = 0$.

Much like their ordinary counterparts, they can be classified based on linearity, order, constant vs. variable coefficients, and their homogeneity.

We can also describe the solutions to partial differential equations. There are more than you can even shake a stick at. For example, consider

$$\frac{\partial u}{\partial t} = \frac{\partial^2 u}{\partial x^2} \tag{9.2}$$

There are many different solutions that are possible.

$$u = \frac{1}{2}x^2 + t \tag{9.3}$$

$$u = e^x e^t \tag{9.4}$$

For this reason, we have to use auxiliary conditions to dictate which solutions are valid.

Initial Conditions

Initial conditions specify the property of $u(x,t)$ at a specific coordinate, $u(x_0,t)$ or at a specific time $u(x,t_0)$.

They may also specify the time derivatives of u at a coordinate or time.

The number of initial conditions that are required is equal to the order of the highest time derivative.

Another auxiliary condition we can consider is the spatial behavior of the function at a time or point. These are known as boundary conditions.

Boundary Conditions

Boundary conditions specifies the behavior of the function $u(x,t)$ at a point in space. Generally, the ends of the spacial domain is used.

There are three types of boundary conditions that are used, depending on the problem.

Dirichlet Conditions

Dirichlet conditions are based on the spatial coordinates alone. For example, you may have the boundary condition

$$u(x_0, t) = h_1(t) \tag{9.5}$$

If u had more than one spatial coordinate, you may choose

$$u(x_0, y, t) = M_1(y, t) \tag{9.6}$$

Newmann Conditions

Newmann conditions rely on the spatial derivatives of $u(x,t)$ at a point x_0. They will be given in the form

$$\frac{\partial u}{\partial x}|_{x=x_0} = h_2(t) \tag{9.7}$$

Robin Conditions

Robin boundary conditions are essentially a concatenated version of Dirichlet and Newmann's conditions. It is the linear combination of the derivative and the function itself at a point $x = x_0$.

$$\frac{\partial u}{\partial x}|_{x=x_0} + ku|_{x=x_0} = h_3(t) \tag{9.8}$$

Lastly, to make note, the number of boundary conditions you need is equal to the sum of the orders of the highest spatial derivatives in each coordinate. For example,

$$\partial_t u = \partial_x^2 u \tag{9.9}$$

Requires 2.

$$\partial_t u = \partial_x^2 u + \partial_y^2 u \tag{9.10}$$

Requires 4.

9.2 Method of Characteristics

The method of characteristics is one of the most powerful techniques for solving first-order partial differential equations analytically. It can be used for linear equations, certain non-linear equations, and what are known as quasi-linear equations. This means that the PDE is linear with respect to it's highest derivative only. So

$$\partial_u \partial_{yy}^2 = f(x,y) \tag{9.11}$$

is quasi-linear, but

$$(\partial_x)^2 + \partial_u = f(x,y) \tag{9.12}$$

is completely non-linear.

Method of Characteristics

Given a partial differential equation of the form

$$a(x,y,u)u_x + b(x,y,u)u_y = c(x,y,u) \tag{9.13}$$

The characteristics of this are given by

$$\frac{dx}{dt} = a(x,y,u) \tag{9.14}$$

$$\frac{dy}{dt} = b(x,y,u) \tag{9.15}$$

$$\frac{du}{dt} = c(x,y,u) \tag{9.16}$$

Once we have the characteristics set up, we solve this system of three equations and ultimately attempt to eliminate t.

9.3 Separation of Variables

Separation of variables is an important technique for solving many first, second, and higher order differential equations. In essence, you are splitting a PDE into a system of ODEs. One ODE solves the time component; the rest solve for the spatial coordinates.

This applies to any linear, homogeneous PDE that also has linear, homogeneous boundary conditions. For example

$$(\alpha\frac{\partial u}{\partial x} + \beta u)|_{x=0} = 0 \tag{9.17}$$

$$(\gamma\frac{\partial u}{\partial x} + \delta u)|_{x=L} = 0 \tag{9.18}$$

Would apply.

When we solve problems using this method, we will find individual solutions $u_n(x,t) = X_n(x)T_n(t)$ that are based off of n. Each corresponding value of n will yield a different eigenvalue in your solution. To find the general solution, we sum over n.

$$u(x,t) = \sum_{n=0}^{\infty} a_n u_n(x,t) \tag{9.19}$$

9.4 Lie Symmetry Analysis

Lie symmetry analysis is an important method for solving certain partial differential equations. Sophus Lie came up with the idea of exploiting symmetries (spatial, time, scaling, etc.) to simplify PDEs. A symmetry takes the form

$$X = \xi(x,t,u)\partial_x + \tau(x,t,u)\partial_t + \phi(x,t,u)\partial_u \qquad (9.20)$$

Then, we use this function X and apply the second prolongation to satisfy

$$X^{(2)}(\Delta)\|_{\Delta=0} = 0 \qquad (9.21)$$

Where Δ is the differential equation that you are trying to solve. The formula and process are quite lengthy, but they have been calculated for certain cases.

$$u_t = u_{xx}$$

#	Symmetry	Generator (Vector Field)	Description
1	Time	$\mathbf{v}_1 = \partial_t$	Shifts time
2	Space	$\mathbf{v}_2 = \partial_x$	Shifts space
3	Scaling u	$\mathbf{v}_3 = u\partial_u$	Multiplies solution by constant
4	Galilean	$\mathbf{v}_4 = t\partial_x + x\partial_u$	Moving frame
5	Dilation	$\mathbf{v}_5 = 2t\partial_t + x\partial_x$	Scales time and space
6	Conformal	$\mathbf{v}_6 = t^2\partial_t + tx\partial_x + (x^2 - 2t)u\partial_u$	Advanced dilation symmetry

$$u_{tt} = c^2 u_{xx}$$

#	Symmetry Name	Generator (Vector Field)	Description
1	Time translation	$\mathbf{v}_1 = \partial_t$	Shifts in time
2	Space translation	$\mathbf{v}_2 = \partial_x$	Shifts in space
3	Scaling	$\mathbf{v}_3 = x\partial_x + t\partial_t$	Dilates space and time
4	Lorentz boost	$\mathbf{v}_4 = t\partial_x + \frac{x}{c^2}\partial_t$	Moving reference frame
5	Solution scaling	$\mathbf{v}_5 = u\partial_u$	Multiply solution by constant
6	Extended inversion	$\mathbf{v}_6 = x^2\partial_x + xt\partial_t$	Conformal-type symmetry

9.5 Heat Equation

The heat equation models how temperature evolves in a continuous medium over time. The earliest forms were developed by Joseph Fourier in the early 19th century, leading to the birth of Fourier analysis and mathematical physics.

Let $u(x,t)$ denote the temperature at position x and time t in a thin, homogeneous rod of length L. Assume the rod has uniform thermal properties and is insulated along its sides, so heat flows only along its length.

We consider a small segment of the rod between positions x and $x + \Delta x$. This allows us to analyze the flow of heat in and out of the segment, leading to a local differential equation.

The amount of heat Q in this segment is proportional to its mass and specific heat capacity:

$$Q = c\rho A \Delta x \cdot u(x,t) \tag{9.22}$$

Here: - c is the specific heat capacity (J/kg·K), - ρ is the density (kg/m^3), - A is the cross-sectional area (m^2), - $u(x,t)$ is the temperature.

The rate of change of heat in the segment is then:

$$\frac{dQ}{dt} = c\rho A \Delta x \cdot \frac{\partial u}{\partial t} \tag{9.23}$$

To model heat conduction, we apply Fourier's law: the heat flux (the rate of heat transfer per unit area) is proportional to the negative of the temperature gradient:

$$\text{Heat flux} = -k\frac{\partial u}{\partial x} \tag{9.24}$$

Here, k is the thermal conductivity (W/m·K). This means heat flows from hotter to colder regions.
The net heat entering the segment is the difference in flux at the boundaries:

$$\text{Net heat in} = A\left[-k\frac{\partial u}{\partial x}(x+\Delta x,t) + k\frac{\partial u}{\partial x}(x,t)\right] \tag{9.25}$$

This simplifies to:

$$\text{Net heat in} = -Ak\left(\frac{\partial u}{\partial x}(x+\Delta x,t) - \frac{\partial u}{\partial x}(x,t)\right) \tag{9.26}$$

Using a Taylor expansion:

$$\text{Net heat in} \approx -Ak\cdot\frac{\partial^2 u}{\partial x^2}(x,t)\cdot\Delta x \tag{9.27}$$

This is the amount of heat added to the segment per unit of time. Equating this to the rate of increase in thermal energy gives:

$$c\rho A\Delta x\cdot\frac{\partial u}{\partial t} = -Ak\cdot\frac{\partial^2 u}{\partial x^2}(x,t)\cdot\Delta x \tag{9.28}$$

Canceling $A\Delta x$ from both sides:

$$c\rho\frac{\partial u}{\partial t} = k\frac{\partial^2 u}{\partial x^2} \tag{9.29}$$

We define the thermal diffusivity α as:

$$\alpha = \frac{k}{c\rho} \tag{9.30}$$

This gives the canonical form of the one-dimensional heat equation:

$$\frac{\partial u}{\partial t} = \alpha \frac{\partial^2 u}{\partial x^2} \tag{9.31}$$

This equation expresses that the rate of change of temperature is proportional to the curvature (second derivative) of the temperature profile. Where the curvature is positive, heat flows in and temperature rises; where it is negative, heat flows out and temperature falls.

9.6 Wave Equation

The wave equation is one of the foundational partial differential equations that one learns about in mathematics and physics. For a wave, it models how it propagates through media. Almost any harmonic system can be related to and converted into the wave equation. In this section, we seek to derive the one-dimensional wave equation for a vibrating string under tension.

Let $u(x,t)$ be the vertical displacement of the string at point x and time t. Assume the string has a uniform mass per unit length ρ, is perfectly flexible, and is stretched under a constant tension T. The goal is to derive an equation that describes the evolution of $u(x,t)$.

We analyze a small element of the string of length Δx, extending from x to $x+\Delta x$. This allows us to apply Newton's laws locally and then generalize using limits.

$$\Delta m = \rho \Delta x \tag{9.32}$$

The vertical motion of this string segment is governed by Newton's second law, $F = ma$. The vertical acceleration is $\frac{\partial^2 u}{\partial t^2}$, so the inertial force is:

$$F = \rho \Delta x \cdot \frac{\partial^2 u}{\partial t^2} \tag{9.33}$$

We now determine the net vertical force on this segment. Since tension acts along the tangent to the string, we consider the vertical components of tension at each end of the segment. For small displacements, the angle of the string is small, and we approximate $\sin\theta \approx \tan\theta$, which is the slope:

$$\sin(\theta) \approx \frac{\partial u}{\partial x} \tag{9.34}$$

This approximation is justified because we assume that the string's oscillations are small, which is typical for linear wave models.

The vertical component of the tension at the left endpoint x is $T\frac{\partial u}{\partial x}(x,t)$, and at the right endpoint $x+\Delta x$, it is $T\frac{\partial u}{\partial x}(x+\Delta x,t)$. The net vertical force is the difference:

$$F_y \approx T\left(\frac{\partial u}{\partial x}(x+\Delta x,t) - \frac{\partial u}{\partial x}(x,t)\right) \tag{9.35}$$

This difference is approximately the derivative of the slope—i.e., the second spatial derivative of the displacement—multiplied by the segment length:

$$F_y \approx T \cdot \frac{\partial^2 u}{\partial x^2}(x,t) \cdot \Delta x \tag{9.36}$$

We now equate the net vertical force with the inertial force using Newton's second law:

$$\rho\Delta x \cdot \frac{\partial^2 u}{\partial t^2} = T \cdot \frac{\partial^2 u}{\partial x^2} \cdot \Delta x \tag{9.37}$$

Dividing both sides by Δx (which is nonzero) removes the dependency on the size of the segment, allowing us to describe motion at any point on the string:

$$\rho \frac{\partial^2 u}{\partial t^2} = T \frac{\partial^2 u}{\partial x^2} \tag{9.38}$$

We now introduce a constant c, representing the speed of wave propagation, defined by:

$$c = \sqrt{\frac{T}{\rho}} \tag{9.39}$$

This allows us to write the final form of the wave equation as:

$$\frac{\partial^2 u}{\partial t^2} = c^2 \frac{\partial^2 u}{\partial x^2} \tag{9.40}$$

This equation describes the motion of transverse waves on a taut string. It reveals that the acceleration of a point on the string is proportional to the local curvature of the string—an intuitive result consistent with the restoring effect of tension.

9.7 Examples

Example 1

Solve the Transport Equation.

$$u_t + cu_x = 0 \qquad u(x,0) = f(x) \tag{9.41}$$

This is set up for the method of characteristics, so we write $\frac{du}{dt}$.

$$\frac{du}{dt} = u_t + \frac{dx}{dt}u_x \qquad u(x,0) = f(x) \tag{9.42}$$

So, we can write the system of ordinary differential equations

$$\frac{du}{dt} = 0 \Rightarrow u = u_0 = u(x_0, t_0) \tag{9.43}$$

$$\frac{dx}{dt} = c \tag{9.44}$$

So, solving both of these, we get the characteristic line

$$x = ct + x_0 \tag{9.45}$$

Along the characteristic, u is constant. It passes through both $u(x,0) = f(x)$ and $u(x,t)$. And, along with the characteristic, if $t = 0$ we have in the initial condition, then $x = x_0$. This means that $u(x,0) = f(x) = f(x_0)$. Since u is constant along the characteristic $u(x,0) = u(x,t)$, $u(x,t) = f(x_0)$. Solving for x_0,

$$x_0 = x - ct \tag{9.46}$$

Thus, the solution to the transport equation is

$$u(x,t) = f(x - ct) \tag{9.47}$$

Example 2

Solve

$$u_x + yu_y = 0 \qquad u(x, 1) = \sin(x) \tag{9.48}$$

Once again, this is set up well for the method of characteristics. In this case,

$$\frac{dx}{dt} = 1 \Rightarrow x = t + x_0 \tag{9.49}$$

$$\frac{dy}{dt} = y \Rightarrow y = y_0 e^t \tag{9.50}$$

$$\frac{du}{dt} = 0 \Rightarrow u = u_0 = u(x_0, y_0) \tag{9.51}$$

Now, to break this down, we first solve for t from the solution to the x differential equation.

$$t = x - x_0 \tag{9.52}$$

Plugging this into y, we get

$$y = y_0 e^{x - x_0} \tag{9.53}$$

We can solve for x_0 from here.

$$x_0 = x - \ln(\frac{y}{y_0}) \tag{9.54}$$

Applying our initial conditions $u(x, 1) = \sin x = \sin x_0$, we get

$$1 = y_0 e^t \tag{9.55}$$

Since $u(x, y_0) = u(x, 1)$ along the characteristic, $y_0 = 1$.

$$t = \ln(\frac{y_0}{y}) \Rightarrow t = -\ln y \tag{9.56}$$

From above,

$$x_0 = x - \ln(\frac{y}{y_0}) \tag{9.57}$$

Using $y_0 = 1$,

$$x_0 = x - \ln y \tag{9.58}$$

Combining this with $u(x, y) = \sin(x_0)$, we obtain the final solution.

$$u(x, y) = \sin(x - \ln y) \tag{9.59}$$

Example 3

Solve the Transport Equation.

$$u_t + cu_x = 0 \qquad u(x,0) = f(x) \tag{9.60}$$

Setting up the method of characteristics, we obtain

$$\frac{dx}{dt} = 1 \Rightarrow x = t + x_0 \tag{9.61}$$

$$\frac{dy}{dt} = u \Rightarrow y = ut + y_0 \tag{9.62}$$

$$\frac{du}{dt} = 0 \Rightarrow u = u_0 = f(x_0) \tag{9.63}$$

Now, if $y = 0$, $y_0 = 0$, and $x - x_0 = t \Rightarrow y = u(x - x_0) + y_0$, we can solve for x_0.

$$x_0 = x - \frac{y}{u_0} \tag{9.64}$$

Using the fact that

$$u(x,y) = f(x_0) \tag{9.65}$$

along the characteristic, we find the final solution.

$$u = f(x - \frac{y}{u}) \tag{9.66}$$

Example 4

Solve the heat equation using Lie symmetries

$$\frac{\partial u}{\partial t} = \frac{\partial^2 u}{\partial x^2} \tag{9.67}$$

To solve using a Lie symmetry, we will attempt scaling.

$$X = 4t\partial_t + 2x\partial_x - u\partial_u \tag{9.68}$$

We can use the method of characteristics to find our invariant transformation in terms of x and t.

$$\frac{dt}{4t} = \frac{dx}{2x} \Rightarrow \frac{dt}{2t} = \frac{dx}{x} \tag{9.69}$$

Solving this by integrating, we get

$$\ln(t^2) = \ln(x) \Rightarrow \frac{x^2}{t} = \text{constant} \tag{9.70}$$

You can try applying this; however, you will find that the linear terms of x are much easier to handle. Despite this being an invariant, we make use of an equivalent invariant.

$$\xi = \frac{x}{\sqrt{t}} \tag{9.71}$$

Now, we can create an ansatz that revolves around this invariant. We will try

$$u(x,t) = t^{\alpha} v(\xi) \tag{9.72}$$

We can take the first time derivative and the second spatial derivatives of this. A useful tip is that

$$\frac{d\xi}{dt} = \frac{\xi}{2t} \tag{9.73}$$

$$u_t = \alpha t^{\alpha-1} v(\xi) + t^{\alpha} v'(\xi) \frac{d\xi}{dt} \tag{9.74}$$

Using the derivative from above, we simplify this to

$$u_t = t^{\alpha-1}(\alpha v - \frac{\xi}{2} v') \tag{9.75}$$

In the same manner, we compute u_{xx}.

$$u_x = t^{\alpha-\frac{1}{2}} v'(\xi) \tag{9.76}$$

$$u_{xx} = t^{\alpha-1} v''(\xi) \tag{9.77}$$

Now, we can plug this into the original heat equation.

$$t^{\alpha-1}(\alpha v - \frac{\xi}{2} v') = t^{\alpha-1} v''(\xi) \tag{9.78}$$

Cancel out $t^{\alpha-1}$ on both sides, leaving

$$(\alpha v - \frac{\xi}{2} v') = v'' \tag{9.79}$$

For the heat equation, we often choose $\alpha = -\frac{1}{2}$

$$v'' + \frac{\xi}{2} v' + \frac{1}{2} v = 0 \tag{9.80}$$

We solve this for $v(\xi)$. It can be expressed as a Hermite Polynomial or simplified using the ansatz $v(\xi) = Ae^{-a\xi^2}$. This matches what we would find by solving the heat equation using a different method, so we can try to verify that our ansatz works.

$$v' = -2a\xi Ae^{-a\xi^2} \tag{9.81}$$

$$v'' = Ae^{-a\xi^2}(4a^2\xi^2 - 2a) \tag{9.82}$$

Substitute everything into the ODE.

$$(Ae^{-a\xi^2}(4a^2\xi^2 - 2a)) + \frac{1}{2}(-2a\xi Ae^{-a\xi^2}) + \frac{1}{2}Ae^{-a\xi^2} = 0 \tag{9.83}$$

From this equation, we can factor out and cancel an $Ae^{-a\xi^2}$ term. We can safely do this since the exponential will never be equal to 0.

$$(4a^2\xi^2 - 2a - a\xi^2 + \frac{1}{2}) = 0 \tag{9.84}$$

We group the coefficients based on the power of ξ.

$$(4a^2 - a)\xi^2 + (-2a + \frac{1}{2}) = 0 \tag{9.85}$$

For this equation to hold, both sets of coefficients must be equal to 0. So,

$$4a^2 - a = 0 \Rightarrow a = \frac{1}{4} \quad \text{or} \quad a = 0 \tag{9.86}$$

$$(-2a + \frac{1}{2}) = 0 \Rightarrow a = \frac{1}{4} \tag{9.87}$$

Thus, for this equation to hold, $a = \frac{1}{4}$. Further, we write $v(\xi)$ as

$$v(\xi) = Ae^{\frac{-\xi^2}{4}} \tag{9.88}$$

Since $\xi = \frac{x}{\sqrt{t}}$ and $u(x,t) = t^{\alpha}v(\xi)$, the final solution to the heat equation is

$$u(x,t) = t^{-\frac{1}{2}} A e^{\frac{-x^2}{4t}} \tag{9.89}$$

If we used a domain restriction for boundaries, then we could calculate A.

Example 5

Solve the heat equation using separation of variables.

$$\frac{\partial u}{\partial t} = \frac{\partial^2 u}{\partial x^2} \tag{9.90}$$

With two boundary and one initial condition.

$$u(0,t) = u(L,t) = 0 \qquad u(x,0) = \phi(x) \tag{9.91}$$

We make the solution

$$u(x,t) = X(x)T(t) \tag{9.92}$$

Substituting this into the PDE, we get

$$\frac{\partial u}{\partial t} = X(x)\frac{dT}{dt} \tag{9.93}$$

$$\frac{\partial^2 u}{\partial x^2} = \frac{d^2 X}{dx^2}T(t) \tag{9.94}$$

Now, we equate these two and move the T to its respective side and the X to its side.

$$\frac{1}{T}\frac{dT}{dt} = \frac{1}{X}\frac{d^2}{dx^2} \tag{9.95}$$

Since the left-hand side is only a function of T and the right-hand side of X, they must equate to one another by a constant, λ. For simplicity, we choose λ^2. This constant can either be positive, negative, or zero. You can show that the positive case does not work because $\lim_{t\to\infty} u(x,t) = \infty$. This is intuitively an issue because heat is not generated out of nothing; that would violate the conservation of energy. For a different reason, the constant case also fails. It would grant a solution of the form

$$u(x,t) = A(Bx + C) \tag{9.96}$$

If we take $u(0,t)$, we should get 0; however, instead we find that

$$AC = 0 \tag{9.97}$$

This only works if A or C is 0, but that would never happen. Thus, it must be the negative case that finally works. This gives us the system of ODEs,

$$\frac{d^2X}{dx^2} = -X\lambda^2 \Rightarrow X(x) = B\cos(\lambda x) + C\sin(\lambda x) \tag{9.98}$$

$$\frac{dT}{dt} = -T\lambda^2 \Rightarrow T(t) = Ae^{-\lambda^2 t} \tag{9.99}$$

If we multiply these together, we get

$$u_n(x,t) = e^{-\lambda^2 t}(B\cos(\lambda x) + C\sin(\lambda x)) \tag{9.100}$$

Now, apply the boundary conditions.

$$u = 0, x = 0 \Rightarrow B = 0 \tag{9.101}$$

$$u = 0, x = L \Rightarrow \sin(\lambda L) = 0 \tag{9.102}$$

Therefore, the eigenvalues λ_n are given by

$$\lambda_n = \frac{n\pi}{L} \qquad \text{for} \quad n = 1, 2, 3, ... \tag{9.103}$$

We can write u_n.

$$u_n = Ce^{-\lambda_n^2 t}\sin(\lambda_n x) \tag{9.104}$$

Any value of C would satisfy the PDE. Lastly, we rewrite this as an infinite series to obtain the complete solution.

$$u(x,t) = \sum_{n=0}^{\infty} a_n e^{-\lambda_n^2 t}\sin(\lambda_n x) \tag{9.105}$$

Chapter 10

Exercises

10.1 General Problems

1. $y'' + y = 0$
2. $y' = 3y$
3. $x^2y'' - 3xy' + 4y = 0$
4. $y'' - 2y' + y = e^x$
5. $y'' + 9y = 18\cos(3x)$
6. $y' + \frac{y}{x} = x$
7. $xy' + y = \sin x$
8. $y'' - 4y' + 4y = 0$
9. $y' = y^2 - 1$
10. $y'' + y' - 2y = 3e^{-x}$
11. $x^2y'' + xy' - y = 0$

12. $y'' + 4y' + 4y = \sin(2x)$

13. $y' = x^2 + 2x + 1$

14. $y'' + y = \cos x$

15. $x^2y'' + 5xy' + 4y = 0$

16. $y' + y\tan x = \sin x$

17. $y'' - 9y = 0$

18. $y'' + y' = 2e^{-x}$

19. $x^2y'' - xy' + y = 0$

20. $y' = y(1 - y)$

21. $y'' + 16y = 8\sin(4x)$

22. $x^2y'' + 3xy' + y = 0$

23. $y'' - 2y' - 3y = 0$

24. $y' = e^{2x}$

25. $y'' + y = \sin x$

26. $xy' - y = x^2$

27. $y'' + 2y' + 10y = 0$

28. $y' = \frac{x}{1+x^2}$

29. $y'' - 4y = e^{2x}$

30. $y'' + y' = x$

31. $y'' - y' - 6y = 0$

32. $x^2y'' - 2xy' + 2y = 0$

33. $y'' + 9y = \sin(3x)$

34. $y' = \frac{1}{1+x^2}$

35. $y'' - 3y' + 2y = 0$

36. $y'' + y = e^{-x}$

37. $x^2y'' + xy' - 9y = 0$

38. $y'' + 2y' + y = e^{-x}$

39. $y' + y = e^x$

40. $xy'' - y' = 0$

41. $y'' - y = \frac{1}{1+e^x}$

42. $y'' + 4y = \cos(2x)$

43. $y' = y + 1$

44. $x^2y'' - 4xy' + 6y = 0$

45. $y'' - 2y' + 2y = 0$

46. $y'' + 9y = e^{3x}$

47. $y' = \sin x$

48. $y'' - 3y' + 2y = e^{2x}$

49. $xy'' + 2y' = 0$

50. $y'' - 4y = \cosh(2x)$

51. $y'' + y = \sin x + \cos x$

52. $y' = \frac{1}{x+y}$

53. $y'' - y = e^x$

54. $x^2y'' - 2xy' + 2y = 0$

55. $y'' + 4y = 8\sin(2x)$

56. $y' + 2y = 3e^{-2x}$

57. $y'' - y' = 0$

58. $y'' + y' + y = 0$

59. $y' = \cos x$

60. $y'' + 4y' + 4y = e^{-2x}$

61. $x^2y'' + 3xy' - y = 0$

62. $y'' + y = x^2$

63. $y'' - 4y = x$

64. $y'' - y' + y = 0$

65. $x^2y'' - xy' + y = 0$

66. $y'' + 25y = 50\cos(5x)$

67. $y' = e^x \sin x$

68. $y'' - 2y' + y = x^2$

69. $x^2y'' + 2xy' + y = 0$

70. $y'' + y' - 2y = e^x$

71. $y' = xe^x$

72. $y'' - 9y = 9e^{3x}$

73. $x^2y'' - 4xy' + 6y = 0$

74. $y'' + y = \tan x$

75. $y'' + 2y' + y = 3$

76. $y' + y = \sin x$

77. $y'' - 5y' + 6y = 0$

78. $x^2y'' + 5xy' + 9y = 0$

79. $y' = \frac{y}{x} + 1$

80. $y'' - y = \sin x$

81. $xy'' - y' = 2x$

82. $y'' + 9y = \sin(3x) + e^{-x}$

83. $y' = \frac{x+y}{x}$

84. $y'' + y' = \cos x$

85. $x^2y'' - 3xy' + 3y = 0$

86. $y'' - y' - 2y = e^{2x}$

87. $y' = \ln x$

88. $y'' + 4y' + 4y = 4e^{-2x}$

89. $y'' + 16y = \sin(4x)$

90. $y'' - 2y' + y = e^x$

91. $x^2y'' + xy' - 4y = 0$

92. $y'' + y' = e^x$

93. $y' = \frac{x^2}{1+x^2}$

94. $y'' - 6y' + 9y = 0$

95. $y'' + y = \frac{1}{1+x^2}$

96. $x^2y'' - xy' + y = 0$

97. $y'' + y' + y = e^{-x}$

98. $y' + 3y = e^{3x}$

99. $y'' - y = \ln x$

100. $y'' + 2y' + 2y = \sin x$

101. $y'' - 2y' + y = \cosh x$

10.2 Word Problems

1. A cup of coffee at 90°C cools in a 20°C room. After 5 minutes, it's 70°C. Use Newton's Law of Cooling to model and predict its temperature after 10 minutes:

$$\frac{dT}{dt} = -k(T - 20)$$

2. A tank contains 100 L of water with 20 g of salt. Brine with 2 g/L salt flows in at 5 L/min, and the mixture flows out at the same rate. Find the salt amount after 10 minutes:

$$\frac{dQ}{dt} = 10 - \frac{5Q}{100}$$

3. The population of a bacteria colony grows at a rate proportional to its size, but limited by a carrying capacity of 10,000. If it doubles from 1,000 to 2,000 in 3 hours, find its growth law:

$$\frac{dP}{dt} = kP\left(1 - \frac{P}{10{,}000}\right)$$

4. A skydiver of mass 75 kg falls vertically with air resistance proportional to velocity. If terminal velocity is 60 m/s, find their velocity as a function of time:

$$m\frac{dv}{dt} = mg - kv$$

5. A radioactive isotope decays at a rate proportional to its remaining mass. If 80% remains after 6 hours, find

the half-life and remaining fraction after 24 hours:

$$\frac{dM}{dt} = -kM$$

6. A spring-mass system has mass 2 kg and spring constant 8 N/m. If it's pulled 0.5 m from equilibrium and released, find the motion assuming no damping:

$$2y'' + 8y = 0$$

7. A car's brakes cause a deceleration proportional to its speed. If the car is moving at 30 m/s and comes to rest after 10 seconds, find its speed as a function of time:

$$\frac{dv}{dt} = -kv$$

8. A tank with 200 L of pure water receives inflow of salt water (1 g/L) at 4 L/min; the mixture drains at the same rate. Find the salt concentration at any time:

$$\frac{dQ}{dt} = 4 - \frac{4Q}{200}$$

9. The temperature of a metal rod satisfies the 1D heat equation. Initially, it's 100°C at one end and 0°C at the other. Describe how the temperature changes over time:

$$\frac{\partial u}{\partial t} = \alpha \frac{\partial^2 u}{\partial x^2}$$

10. A circuit has an inductor (2 H), resistor (4), and battery (12 V). Find the current over time after the

switch is closed, assuming initially $I(0) = 0$:

$$2\frac{di}{dt} + 4i = 12$$

11. A tumor grows at a rate proportional to its current size but slows as it approaches a maximum mass of 200 g. If it's 10 g initially and 15 g after a week, find its mass over time:

$$\frac{dM}{dt} = kM\left(1 - \frac{M}{200}\right)$$

12. A parachutist of mass 70 kg jumps from 1,000 m. Assuming air resistance $F = kv^2$, find the time to reach 90% of terminal velocity:

$$m\frac{dv}{dt} = mg - kv^2$$

13. A chemical reaction $A \to B$ follows first-order kinetics with rate constant k. If 25% of A remains after 30 minutes, find k and the amount after 1 hour:

$$\frac{d[A]}{dt} = -k[A]$$

14. The motion of a damped spring is given by $my'' + cy' + ky = 0$. A 1 kg mass attached to a spring (k = 9 N/m, c = 3 Ns/m) is displaced 0.2 m and released. Find its motion:

$$y'' + 3y' + 9y = 0$$

15. The rate of change of a drug in the bloodstream is proportional to the amount present. If a 50 mg dose drops to 25 mg after 4 hours, how long until only 10 mg remains?
$$\frac{dC}{dt} = -kC$$

16. The spread of a rumor in a population is proportional to both the number who've heard it and those who haven't. Write a model for this process:
$$\frac{dR}{dt} = kR\left(1 - \frac{R}{N}\right)$$

17. A tank leaks water at a rate proportional to the square root of its depth. If the tank initially holds 1 m of water, how long until it is half-empty?
$$\frac{dh}{dt} = -k\sqrt{h}$$

18. An investment grows at 5% continuously compounded interest, but withdrawals occur at \$100 per year. If the account starts at \$5,000, find the balance over time:
$$\frac{dA}{dt} = 0.05A - 100$$

19. A heated object is placed in a room that oscillates in temperature between 15°C and 25°C every hour. Find the long-term temperature behavior of the object:
$$\frac{dT}{dt} = -k(T - (20 + 5\sin(\omega t)))$$

20. A disease spreads in a city according to $\frac{dI}{dt} = kI(1 - I/N)$, where $N = 10{,}000$. If 100 are infected initially and 500 after 5 days, find k:

$$\frac{dI}{dt} = kI\left(1 - \frac{I}{10{,}000}\right)$$

21. The rate at which snow melts is proportional to the difference between 0°C and the air temperature. If it takes 3 hours to melt half the snow, predict how long until 90% is gone:

$$\frac{dS}{dt} = -k(0 - T_{\text{air}})$$

22. The motion of a pendulum of length L is governed by $\theta'' + \frac{g}{L}\sin\theta = 0$. Linearize for small oscillations and find the approximate period:

$$\theta'' + \frac{g}{L}\theta = 0$$

23. A 20 kg block slides down a frictional incline where frictional force equals kv. If $k = 5$, find the terminal velocity:

$$m\frac{dv}{dt} = mg\sin\theta - kv$$

24. A mixing tank initially contains 50 L of a 20% alcohol solution. Water enters at 5 L/min, and the mixture drains at 5 L/min. Find the concentration after 10 minutes:

$$\frac{dQ}{dt} = -\frac{5Q}{50}$$

25. A spaceship in deep space expels fuel at a constant rate, producing a thrust proportional to the exhaust velocity. Write the ODE describing the velocity as mass decreases:

$$m\frac{dv}{dt} = -v_e\frac{dm}{dt}$$

26. A charging capacitor in an RC circuit (R = 1 k, C = 10 µF) is connected to a 12 V source. Find the voltage across the capacitor as a function of time:

$$RC\frac{dV_c}{dt} + V_c = 12$$

27. A predator-prey system follows $\frac{dx}{dt} = x(1-y)$, $\frac{dy}{dt} = y(x-1)$. Describe the equilibrium points and local behavior:

$$\begin{cases} \frac{dx}{dt} = x(1-y) \\ \frac{dy}{dt} = y(x-1) \end{cases}$$

28. A population of fish grows logistically but is harvested at a constant rate H. Find the condition under which the population stabilizes or collapses:

$$\frac{dP}{dt} = rP\left(1 - \frac{P}{K}\right) - H$$

29. A car's suspension system can be modeled by $my'' + cy' + ky = F_0\cos(\omega t)$. Find the steady-state amplitude for a given driving frequency:

$$my'' + cy' + ky = F_0\cos(\omega t)$$

30. A tank of 500 L contains water with 10 kg of dissolved sugar. Fresh water flows in at 5 L/min, and mixture flows out at the same rate. Find the sugar amount after 15 minutes:

$$\frac{dQ}{dt} = -\frac{5Q}{500}$$

Appendix A

Existence and Uniqueness Theorems

Existence and uniqueness theorems are important when studying differential equations because they tell us whether we are guaranteed a solution given a certain problem. This is vital, theoretically, to understanding them and for practical applications.

Existence tells us whether we can find a solution, numerical or analytic, without actually having to find it. Uniqueness theorems tell us whether there is only one solution given a differential equation and initial or boundary conditions.

A.1 Cauchy-Lipschitz Theorem

This theorem implies local existence and uniqueness for certain differential equations.

Cauchy-Lipschitz Theorem

Given the ODE

$$\frac{dy}{dt} = f(t, y), \qquad y(t_0) = y_0 \tag{A.1}$$

If $f(t, y)$ is continuous in a rectangle D around the point $f(t_0, y_0)$ and Lipschitz continuous in y, then there exists a unique local solution passing through (t_0, y_0).

To prove this, we set the initial value problem (IVP) up as an integral equation.

$$y(t) = y_0 + \int_0^t f(s, y(s))\, ds \tag{A.2}$$

We define Picard approximations.

$$y_{k+1}(t) = y_0 + \int_{t_0}^t f(s, y_k(s))\, ds \tag{A.3}$$

Next, we need to show that $y_k(t)$ will converge to a unique function, $y(t)$.

Let $X = C([t_0 - h, t_0 + h], \mathbb{R}^n])$ be the space of continuous functions with the supremum norm

$$\sup_{|t-t_0|\leq h} \|y(t)\| = \|y(t)\| \tag{A.4}$$

This is a Banach space. The corresponding Picard operator is

$$(Ty)(t) = y_0 + \int_0^t f(s, y(s))\, ds \tag{A.5}$$

Then, we show that this is a contraction mapping so that we can apply the Banach Fixed-Point Theorem. Using the Lipschitz condition that we assumed,

$$\|f(t, y_1) - f(t, y_2)\| \leq L\|y_1 - y_2\| \quad \forall \quad (t, y_1), (t, y_2) \in D \tag{A.6}$$

$$\Rightarrow \|Ty_1 - Ty_2\| = \sup_{|t-t_0|\leq h} \| \int_0^t f(s, y_1(s)) - f(s, y_2(s))\, ds\| \tag{A.7}$$

Using the integral triangle inequality,

$$\leq \sup_{|t-t_0|\leq h} \int_0^t \|f(s, y_1(s)) - f(s, y_2(s))\|\, ds \tag{A.8}$$

This must then be less than

$$\sup_{|t-t_0|\leq h} \int_0^t L\|y_1 - y_2\|\, ds \tag{A.9}$$

$$\leq Lh\|y_1 - y_2\| \tag{A.10}$$

We choose h such that

$$Lh < 1 \tag{A.11}$$

which makes T a contraction mapping. Therefore, we are able to apply the Banach Fixed-Point Theorem. A contraction on a Banach space has a unique fixed point $y(t)$ such that

$$Ty = y \Rightarrow y(t) = y_0 + \int_0^t f(s, y(s))\, ds \tag{A.12}$$

Since this is equivalent to our ODE, our solution $y(t)$ must be locally existent and unique.

A.2 Peano Existence Theorem

If we wanted to bypass the use of the Lipschitz condition, then we would have to give up uniqueness. This theorem implies that there is at least one solution passing through the point (t_0, y_0).

Peano Existence Theorem

Given the ODE

$$\frac{dy}{dt} = f(t, y), \qquad y(t_0) = y_0 \tag{A.13}$$

If $f(t, y)$ is continuous in a rectangle R around the point $f(t_0, y_0)$, then there exists at least one local solution passing through (t_0, y_0).

To prove this theorem, we make use of Euler's method. First, note that if $f(t, y)$ is continuous in R, then it is also bounded.

$$\|f(t, y)\| \leq M \tag{A.14}$$

We can use Euler's method and divide the interval $[t_0, t_0+h]$ into n subintervals.

$$t_k = [t_0, t_0 + \frac{h}{n}] \tag{A.15}$$

We define the polygon function $y_n(t)$ recursively.

$$y_n(t_0) = y_0 \tag{A.16}$$

$$y_n(t_{k+1}) = y_n(t_n) + \frac{h}{n} f(t_k, y_n(t_k)) \tag{A.17}$$

Each y_n satisfies

$$\|y_n(t) - y_0(t)\| \leq hM \tag{A.18}$$

This bounds the growth, so the approximations all remain within the rectangle R. The slope of each linear piece of the approximation is also bounded by M. Because of this, the family of functions $\{y_n\}$ is equicontinuous. Now, we can apply the Arzela-Ascoli Theorem. It states that a uniformly bounded, equicontinuous sequence has a uniformly convergent subsequence.

We let $y(t)$ be the uniform limit of $y_n(t)$ as $n \to \infty$. The integral equation using y_n becomes

$$y_n(t) = y_0 + \int_0^t f(s_n, y(s_n))\, ds \tag{A.19}$$

Since, as $n \to \infty$, the step size for the approximation goes to 0, the finite steps converge to the integral itself. Additionally, since we let $y_n(t) \to y(t)$ as $n \to \infty$, the right-hand side will converge as well. So, we have

$$y(t) = y_0 + \int_0^t f(s, y(s))\, ds \tag{A.20}$$

Therefore, there must exist a $y(t)$ that is a solution to the differential equation passing through (t_0, y_0).

Appendix B

Resources

For more problems, lessons, and insights on differential equations, consider any of the following resources:

1. Ordinary Differential Equations by William A. Adkins and Mark G. Davidson
2. A First Course in Differential Equations by J. David Logan
3. Differential Equations by Chris McMullen
4. Introduction to Partial Differential Equations with Applications by E.C. Zachmanoglou and Dale W. Thou
5. Partial Differential Equations (Youtube) by Faculty of Khan

These are the main sources that allowed me to write this book. Credit goes to these authors where it is due.

www.ingramcontent.com/pod-product-compliance
Lightning Source LLC
LaVergne TN
LVHW100303210926
848480LV00003B/1007
* 9 7 9 8 3 4 9 6 5 5 2 2 7 *